THE LITTLE BOOK OF
FUNGI

THE LITTLE BOOK OF
FUNGI

With color illustrations by Tugce Okay

BRITT A. BUNYARD

PRINCETON UNIVERSITY PRESS
PRINCETON AND OXFORD

Published in 2024 by Princeton University Press
41 William Street, Princeton, New Jersey 08540
99 Banbury Road, Oxford OX2 6JX
press.princeton.edu

Library of Congress Control Number 2024932728
ISBN 978-0-691-25988-8
Ebook ISBN 978-0-691-26014-3

Typeset in Calluna and Futura PT

Printed and bound in China
1 3 5 7 9 10 8 6 4 2

British Library Cataloging-in-Publication Data is available

This book was conceived, designed, and produced by UniPress Books Limited

Publisher: Jason Hook
Managing editor: Slav Todorov
Creative director: Alex Coco
Project development and management: Ruth Patrick
Design and art direction: Lindsey Johns
Copy editor: Caroline West
Proofreader: Robin Pridy
Color illustrations: Tugce Okay
Line illustrations: Ian Durneen

IMAGE CREDITS:

Alamy Stock Photo: 31 Hakan Soderholm; 51 Michele Cornelius; 61 Clement
Philippe/Arterra Picture Library; 125 Christian Weinkötz; 132 Steven Gill;
137 Ayhan Özdemir. **Getty Images**: 97 Henry Nicholls. **Nature Picture Library**:
26 Alex Hyde. **Science Photo Library**: 57 Carolina Biological Supply Company;
85 Eye of Science. **Shutterstock**: 53 Liga Petersone; 69 Petar B photography;
71 Henri Koskinen; 73 Mindhive; 89 Kateryna Kon; 108 Rodica Vasiliev; 111 Olha
Rohulya; 131 LFRabanedo. **Other**: 13 Dr Alex Hyatt, CSIRO; 21 Angel-Ros-Die;
39 bjoerns; 44 Oluna & Adolf Ceska (MO Observers); 102 Hamilton (MO
Observer); 114 The Next Gen Scientist; 122 Drew Henderson (MO Observer);
149 Jonathan Frank; 151 Britt Bunyard. **Additional illustration references**: 35 Alan
Rockefeller; 37 Bernard Gagnon; 41 Jerzy Opioła; 67 zaca (MO Observer);
113 Workman; 139 Noah Siegel (MO Observer); 147 Aurora Storlazzi.

Also available in this series:

THE LITTLE BOOK OF
BEETLES

THE LITTLE BOOK OF
BUTTERFLIES

THE LITTLE BOOK OF
DINOSAURS

THE LITTLE BOOK OF
SPIDERS

THE LITTLE BOOK OF
TREES

THE LITTLE BOOK OF
WEATHER

THE LITTLE BOOK OF
WHALES

CONTENTS

Introduction   8

1. What is a Fungus?

Fungi are eukaryotes like us   10
How many fungi are there?   12
Primitive, but highly specialized   14
Morphological classification   16
Cup fungi (ascomycetes)   18
Determining relatedness   20

2. Form and Function

Mushrooms: different forms,
    different functions   22
Bracket fungi (polypores)   24
Puffballs   26
Strange mushrooms (stinkhorns)   28
Prized mushrooms (truffles)   30
Rusts, smuts, and blights   32

3. Habitat and Ecology

Fungi everywhere   34
Deserts   36
Arctic and extreme habitats   38
Fungi in urban areas   40
Fungi in our homes   42
Fire fungi   44

4. Evolution

Fossil record   46
The arrival of fungal groups   48
Convergent evolution   50
Truffles and false truffles   52
Co-evolution and migration   54
One pathogen, different hosts   56

5. Lifestyle and Physiology

Our rotten world   58
Brown rot and white rot   60
Fungi with a unique lifestyle   62
Animal parasites   64
Coprophilous fungi   66
Bioluminescence   68

6. Reproduction

The sex lives of fungi   70
Spore production   72
Passive spore dispersal   74
Asexual oomycetes   76
Active dispersal and zoochory   78
Active dispersal and mimicry   80

7. Mutualism and Competition

Mycorrhizal fungi   82
Endomycorrhizae   84
Lichens   86
Mycotoxins   88
Endophytic fungi   90
Mycorrhizal thieves   92

8. Study and Cultivation

The history of mycology   94
Hidden life   96
The age of citizen science   98
Cultivation   100
Extinction   102
Detecting invasive species   104

9. Fungi and Humans

Fungi as food   106
Sustainable myco materials   108
Fungi as cause and cure
    of disease   110
Fungi as medicine   112
Fungi and popular culture   114
Mycotourism   116

10. Threats and Conservation

A changing climate   118
Invasive species   120
Wildfire threats   122
Habitat loss   124
Invasive fungi as pathogens   126
Microbes for a healthy
    body—and planet   128

11. Popular Culture

Fungi fascinate   130
The most legendary mushroom   132
Mushroom of fables   134
Altering the course of history 136
Necrotrophs and sarcophiles   138
Witchery   140

12. Curious Facts

Predacious fungi   142
The largest fungi   144
The tiniest fungi   146
Fungi in unusual environments   148
Toxic fungi   150
Fungal mimicry   152

Glossary   154
Further Reading   156
Index   158
Acknowledgments   160

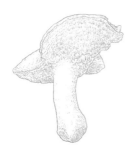

INTRODUCTION

I am indebted to the many educators and mentors who have influenced me throughout my life. I first got interested in wild mushrooms as a kid growing up on a farm in southwestern Ohio. Trying to sort out the fungi found growing in the woods and pastures nearby, the first field guide I ever held was *The Audubon Society Field Guide to North American Mushrooms* by Gary Lincoff. (Decades later, Gary and I would become friends and colleagues!) My head was filled with mushroom names as a student at Kent State University under the instruction of Professor Samuel Mazzer (himself a student of the great Alexander H. Smith).

I received an MS in Plant Biochemistry from Clemson University, studying endophytic fungi and phytohormones, and a PhD in Plant Pathology from Penn State University, where I studied the evolution of macrofungi in the lab of Professor Daniel J. Royse. Following graduation, I was a postdoctoral fellow at the USDA-ARS at Ft. Detrick, Maryland, where I was immersed in the DNA of fungal, oomycete, and bacterial pathogens of potato. I have always had an interest in symbioses—mutually beneficial as well as parasitic and deleterious interactions between fungi and other organisms, be they plants, insects, and even other fungi. The main focus of my research has centered on the co-evolution of macrofungi and Diptera, the true flies.

FUNGI EVERYWHERE

Fungi occur just about everywhere. Try to get to know them—what are they, what are they doing there, and how are they indispensable in that environment? It's what I do! I have had the great fortune to travel all over the world, looking for and marveling at the myriad of fungi I have encountered. A big thanks to the many dozens of mycological societies and mycophiles from across North America (and beyond) who, over the years, have hosted me (and even taken me to their "secret spots"), allowing me to study and photograph so many of the species depicted in this book. This book would not have been possible without their help!

ABOUT THIS BOOK

Besides being a collection of facts that you are likely to read in other mycological resources, this book includes many of my favorite stories about fungi from all over the world. Chapters 1–3 explain what fungi are and how they differ from other eukaryotes. Their form and function are also discussed, as well as their role in the environment. Chapters 4–7 build on this information and explore how it relates to their evolution, physiology, ecological strategies, and how they reproduce. As you will see from the first half of the book, while seemingly simple organisms, fungi are quite different from other life in many ways! Chapters 8–10 in the second half of the book deal with studying fungi, their interactions with us, and threats to fungi, as well as their conservation. Chapters 11 and 12, the final chapters, are just plain fun: myths and folklore about mushrooms and other fungi, fungi in popular culture, and curious facts. I hope you enjoy reading about these fascinating organisms as much as I have enjoyed writing about them.

Britt Bunyard

FUNGI ARE EUKARYOTES LIKE US

All living things are comprised of cells; some organisms are single-celled while many others, like us, are multicellular. All life is divided into prokaryotes, like bacteria, and eukaryotes, which comprise all the rest.

Prokaryotes are single-celled organisms with no membrane-bound organelles or nucleus; DNA consists of a single, circular chromosome. In addition to a cell membrane, bacteria may or may not have a rigid cell wall. Eukaryotes are much more organized physiologically, feature membrane-bound organelles like mitochondria and a nucleus, and DNA is organized as complex chromosomes. Eukaryotes may be single- or multicelled, and include protists, plants, animals, and fungi.

PROTISTS, PLANTS, AND ANIMALS

Protists (kingdom Protista) are single-celled and include an array of forms, many of which you are likely familiar with, such as paramecia, amoebae, and slime molds. Protists may be autotrophic (some are photosynthetic) or heterotrophic: living as saprobes, predators, or parasites. In addition to a cell membrane, protists may or may not have a rigid cell wall; in some cases, the cell wall consists of cellulose (long chains of carbohydrates connected by a specific kind of chemical bond).

All plants have cell walls made of cellulose (and are thus thought to have arisen from a branch of Protista). In addition to other membrane-bound organelles, plant cells have chloroplasts; nearly all plants are autotrophic by means of photosynthesis (a few are parasitic). Animal cells never have cell walls; all animals are heterotrophic.

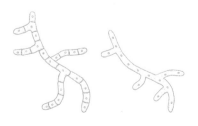

← Fungal cells feature a cellular membrane inside a cell wall made from chitin. Fungal cells form a long, filamentous hypha that elongates at the apical tip.

FUNGUS-LIKE ORGANISMS

Several groups of organisms resemble fungi but can be distinguished on the basis of physiology. Myxomycetes, or slime molds, are so named for their resemblance to fungi, as they ooze along in thin cytoplasmic streams (they are, however, protists). Filamentous algae resemble fungi, although they are photosynthetic and grow by repeated cell divisions to form a chain of cells, unlike fungi, which feature apical growth of hyphal tips. Oomycetes, or water molds, were long thought to be fungi, as their hyphae are all but indistinguishable from those of true fungi. Many of the 500 oomycete species are economically important pathogens of crop plants, trees, and even fish.

FUNGI

All fungi have cell walls made of chitin (somewhat similar to cellulose, but made of long chains of carbohydrates connected by a different specific chemical bond). Fungi derive their energy from all heterotrophic means imaginable—and probably some fungi do things you could not imagine. Many fungi are saprobes, rotting organic matter; probably, most fungi are symbionts, living as mutualistic partners with other organisms or as parasites. It is likely that all plants have species-specific fungal pathogens, including our agricultural crops. A few fungi are carnivorous, trapping and killing their animal prey as a source of nitrogen.

~ The individual fungus ~

Most fungi grow as multicellular filamentous hyphae; a single hypha is a long and thin, tube-like cell, approximately $1/25$ the diameter of a human hair. Some fungi can also grow as a single-celled yeast form; a few fungi seem to favor growing as yeasts. In the vast majority of fungi (indeed, all "higher" fungi), these tubular cells are compartmentalized and divided by septa. As fungi grow, hyphae mass together to form tissue known as a mycelium. Mushrooms like the boletes and chanterelles, as well as tough woody polypores, are made of hyphae.

HOW MANY FUNGI ARE THERE?

There are about 100,000 species of named fungi. It has been estimated that there are probably more like 1.5 million species in total—thus the vast majority of fungi await discovery and description. Fungi are cryptic, their microscopic size (of most) making them difficult to find, and those that elude culture often remain unknown. We do know there are many unseen fungi out there because they leave their DNA behind in soil and other substrates.

SWIMMING FUNGI: THE CHYTRIDS

Long considered the most primitive of the true fungi are the chytridiomycetes. Chytrids are the only motile fungi, producing zoospores that are propelled by flagella. Chytrids feature hyphal growth, although it is not always seen; their hyphae are aseptate. Found worldwide, chytrids are mostly saprotrophs, feeding on decomposing organic matter; some species are parasites of eukaryotes, including plants and animals. Chytrids are linked to the worldwide die-off of amphibians. All fungi placed above chytrids on the fungal tree of life are nonmotile.

COENOCYTIC FUNGI

The zygomycetes were always a mixed bag of fungi placed together by virtue of having coenocytic (aseptate) hyphae. A commonly seen example is the Black Bread Mold (*Rhizopus stolonifer*). Glomeralean fungi were once part of the zygomycetes but have now been elevated to their own phylum, the Glomeromycota. Poorly known, few have been seen or cultured. Few, if any, have sexual reproduction and form no obvious fruiting bodies. Some form clusters of asexual spores, and that's about it. What is known about them is that they are likely the puppet masters of all life on the planet. Glomeralean fungi are mutualistic symbionts of most plants.

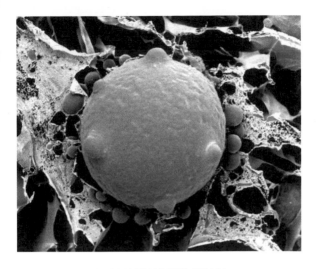

ORGANIZING THE FUNGI

The major groups of fungi have been classified according to characteristics of their sexual reproductive structures. Until recently, fungi were grouped into four classes: chytridiomycetes, zygomycetes, basidiomycetes, and ascomycetes. This taxonomic scheme, although overly simplified, is still a pretty useful system for understanding what these fungi are and how they reproduce. More recently developed classification schemes further separate fungi into additional classes (or phyla); not all scientists agree on the taxonomic hierarchies for some of the oddball groups.

↑ Chytrids are the most primitive fungi and occur worldwide. Most chytrids are saprotrophs, feeding on decomposing matter, but some are serious pathogens of animals.

PRIMITIVE, BUT HIGHLY SPECIALIZED

It is often assumed that organisms more primitive than humans are "lower" or "less evolved." Although appearing less sophisticated, and certainly more ancient evolutionarily, all organisms are highly adapted and specialized for their particular niche.

Fungi have evolved numerous sophisticated methods in order to disperse their reproductive propagules—their spores. Coenocytic fungi do not make complex mushrooms for reproduction, but have their own tricks. Chytrids often parasitize aquatic organisms and have evolved swimming zoospores that seek out their quarry.

Pilobolus—the Hat Thrower—is a zygomycete fungus capable of ejecting spores over great distances. The dung of grazing mammals is a prime habitat for many fungi, and *Pilobolus* is often the first to colonize and the first to sporulate. *Pilobolus* makes sure its spores are already inside the dung when they leave the animal—they are consumed by grazing animals. It launches spore packets, squirting them well outside the "zone of repugnance" and onto clean grass.

↙ A single *Pilobolus* sporangiophore is like a squirt gun only a few millimeters in length, but it can shoot a black sporangium "hat" a distance of 6½ ft (2 m) to a suitable spot where it can be ingested by grazing animals.

→ *Pilobolus* sporangiophores are bulbous-shaped. Each one acts as a lens to focus sunlight, causing the sporangiophore to throw its "hat" into a clearing and away from the dung substrate. The black sporangium is a packet of spores, built to withstand the digestive enzymes of herbivorous mammals.

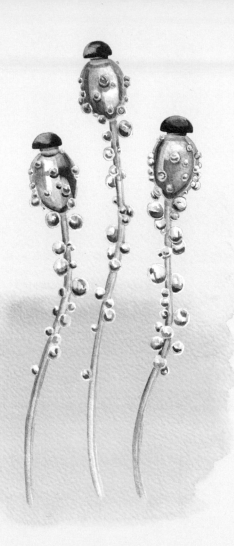

MORPHOLOGICAL CLASSIFICATION

Basidiomycete and ascomycete fungi are collectively known as the "higher" fungi. Other than mycologists, most people are only familiar with the larger, showy fungi of the basidiomycetes and a few ascomycetes. All morphological features of the mushroom, including the color of the spores (and the spore shape, which requires a microscope to see), are useful in determining a fungus species. The basidiomycetes, with their familiar mushroom cap and stalk, are described here.

THE MUSHROOM CAP

With basidiomycete fungi, the mushroom cap (or pileus) is the top part and where the spores are produced. Turn the mushroom over and on the underside of the cap either gills, pores, or spines will be seen. Gills appear as long openings from the edge of the cap to the stalk and are separated by a thin tissue. Pores, on the other hand, are like long tubes running through the underside of the fruiting body and will appear as small holes. The spores are released from these holes. Spines (or "teeth") hang from the underside of the cap. The spores are formed on the outside of the spines.

THE MUSHROOM STALK

The stalk is the structure that supports the cap and can be long or very short—or there can be none at all. Mushrooms without stalks are called sessile. Stalks don't always have to be in the center of the cap; they can be off-centered (or eccentric). The bottom of the stalk, at ground level or below, can be round like a bulb or straight. When cut open, the inside can be firm, hollow, or spongy. Often there will be thin tissue covering the gills of the immature mushroom. As the cap matures and expands, the veil breaks and remnants of the partial veil may remain as a ring (or annulus) on the stalk.

THE SPORES

The whole point of this structure is for reproduction, and the end result is spores, the tiny, dust-like reproductive propagules of the fungus. Spores are borne on the walls of the gills, teeth, or pores of the underside of the mushroom cap. Spore color may vary from one group of mushrooms to another, and can be helpful in identifying mushrooms. Some colors are common for many groups, especially white or brown. In contrast, only a few groups of mushrooms have pink spores, and only one species has green. To check the color of a mushroom's spores, you can make a spore print. Simply place a fruiting body on paper (or suspend it just above), cover it with a bowl (creating a chamber that will keep the mushroom humid), and leave for a few hours or overnight.

THE "HUMONGOUS FUNGUS"

What we think of as mushrooms are, in fact, only the fruiting bodies of a much larger organism. Like icebergs, the bulk remains hidden from view, as an underground cobweb of colorless, near-invisible, hair-like filaments called hyphae. Each fungus possesses a dense, intertwining network of hyphae, called the mycelium. The mycelium originates from a spore, and, given enough time, can group to humongous proportions. The largest organism on this planet is an individual specimen of the honey fungus *Armillaria mellea* in North America, whose mycelial mat covers a staggering 2,384 acres (965 hectares) of land in Oregon's Blue Mountains, although its fruiting bodies are no larger than grocery store button mushrooms.

↘ The mushroom is the reproductive part of many kinds of fungi. Turn the mushroom over and on the underside of the cap either gills, spines, or pores—as with these boletes—will be seen. Pores are like long tubes running through the underside of the fruiting body and will appear as small holes.

CUP FUNGI (ASCOMYCETES)

Ascomycetes are the largest group of fungi and produce sexual spores in a special sac-like structure called an ascus. Ascomycetes are the most diverse group, living as saprotrophs, parasites, and mutualistic symbionts (including mycorrhizas and lichens). This group includes morels, truffles, and yeasts.

Ascomycete fungi release spores very differently from basidiomycetes in a way often likened to a squirt gun. Spores are formed within elongate pouches called asci. Asci line the surfaces of cup mushrooms or within chambers like perithecia, themselves hidden within stromal tissue (a compact mass of tissue). As the fruiting body matures, liquid flows into the ascus, causing it to swell. Building pressure causes the ascus tip to rupture and ascospores are ejected.

With some large cup fungi, spore release can be a puff that is easily seen and heard! Once a fruiting body is mature and the asci ready to fire, a simple disturbance of air may be all that is necessary to get them to discharge simultaneously.

↓ Ascospore production: nuclei fuse in the hyphal tip, shown as an elongate pouch called an ascus. Asci line the surfaces of cup mushrooms or within chambers like perithecia (far left); when mature, spores are ejected (far right).

→ The morel mushroom, which is created for ascomycete reproduction, is prized as a gourmet delicacy. Morels are found all over the world and anticipated by impassioned pickers who eagerly await their springtime emergence.

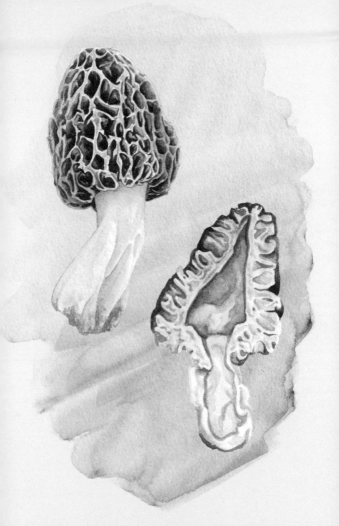

DETERMINING RELATEDNESS

While delightful to look at, plants produce flowers for a purpose: reproduction. More specifically, sexual reproduction. And it makes logical—as well as evolutionary—sense that closely related species produce similar kinds of flowers. Just as flowers are the basis for the classification of plants, so the spore-bearing structures of fungi are used in their classification.

ASEXUAL FUNGI:
TAXONOMICALLY PROBLEMATIC

While we know that fungi often reproduce sexually—we can see their reproductive structures (mushrooms)—we also know that those very same species also sometimes, and maybe even most of the time, reproduce asexually. This can be as simple as the fragmentation of hyphae or the production of asexual spores (clones of the parent), which are known as conidia. If fungi are classified on the basis of how they reproduce sexually, then what to do with asexual forms?

The sexual form of fungus is known as the teleomorphic life cycle state, while the asexual form is referred to as the anamorphic one. A great many important fungi are known only as anamorphs. Many of these are economically important and cause damage to our crops, rot our stored foods, or cause mycoses. Anamorphic fungi are troublesome for another reason—at least to the taxonomists whose job it is to come up with names for them.

~ DNA sequence analysis to the rescue ~

These "imperfect" (asexual) fungi were historically lumped into one big group (the deuteromycetes or *fungi imperfecti*), regardless of evolutionary relatedness. With DNA sequence analysis, researchers could at long last finally determine teleomorphic states, and thus teleomorphic names, for any fungus without the need to produce sexual spores in culture.

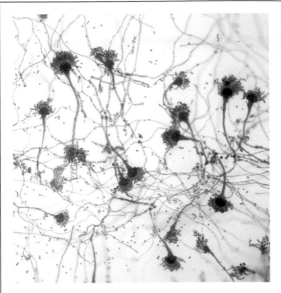

DNA ANALYSIS LEADS TO
SOME SURPRISES!

A common mold, *Aspergillus*, is a well-known asexual fungus. DNA analysis found this fungus has many different sexually reproducing forms—in no fewer than 11 teleomorphic genera! Well, in 2012 scientists had to change the rules on how things are named, making allowances for well-established asexual names in cases where going to the sexual name would be a major headache. Thus, for some *Aspergillus* species (including notorious mycotoxin producers like *Aspergillus flavus* and *A. parasiticus*), the anamorphic name is retained.

MUSHROOMS: DIFFERENT FORMS, DIFFERENT FUNCTIONS

T he basidiomycetes include most of the mushrooms familiar to everyone. They are also called club fungi because they produce sexual spores on club-like stalks called basidia. Mushrooms are the reproductive structures and the basidia are found on the fertile surface, called the hymenium.

The most familiar types of mushrooms are agarics and boletes. Both have a stalk with a cap atop, but the difference lies underneath the cap. Agarics mostly feature a hymenium of gills (or lamellae). Boletes look similar but have a hymenium underneath the cap with what appear to be pores. In reality, those pores are the openings of many tubes from where the spores arise.

There are many other mushroom forms in nature, each one highly adapted to producing spores in that particular niche by a given species of fungus. Bracket fungi live inside a host tree and when it is time to reproduce, they produce a shelf-like mushroom right off the side of the tree (see pages 24–25 for more on bracket fungi).

↓ Beneath the mushroom cap you will find either gills, pores, or spines, and this hymenial surface is where the spores are produced and released.

↓ Lining the hymenial surface are specialized hyphal tips called basidia, which have pointed outgrowths known as sterigma, where the spores will then develop.

→ Vibrant golden *Laetiporus sulphureus* is a highly prized polypore, commonly known as Chicken of the Woods due to its texture and flavor. It is a wood rot fungus that lives inside its host all year round. When it is time to reproduce, overlapping shelves emerge; spores are released from the numerous pores on the underside of each shelf.

BRACKET FUNGI (POLYPORES)

Bracket and shelf fungi don't usually look like much but can be rather interesting. They often have unique physiologies and many are perennial, persisting on their woody hosts year-round, so you can observe them in the middle of winter. Many polypores are thought to be pathogens and are commonly found growing from the main stems of living trees. More often than not, they are saprobes, restricted to rotting the dead heartwood of the tree. Many continue to grow on wood long after the host tree has died. They can also grow to enormous sizes, often larger than a dinner plate.

THE TINDER POLYPORE

Fomes fomentarius is one such polypore. Known as the Tinder Polypore, this cosmopolitan fungus produces large brackets that are commonly seen throughout the Northern Hemisphere. The brackets have been used for fire-starting, probably since the Paleolithic Era, 15,000 years ago. During excavations of prehistoric villages in Italy and Switzerland, the remains of *F. fomentarius* confirm its use for kindling fires. In Germany and France, a cottage industry was based on the manufacture of fire-starter kits utilizing this fungus. Each kit included prepared tinder fungus, a striking steel, and a shaped silica stone, all packaged in a small tin box or a little bag.

← The polypore *Fomes fomentarius* is known as the Tinder Polypore, but it also goes by the name of Horse Hoof Polypore due to its hoof-like appearance that results from perennial layers of growth.

THE SPORE EATER

To the casual observer, *Sporophagomyces chrysostomus* appears to be a dirty whitish to brownish mold growing on the underside of Artist's Conk (*Ganoderma* species) or other woody polypores. But this fungus is probably neither saprobe nor parasite. As the name implies, *Sporophagomyces* is an eater of spores. You might imagine that this is an unusual lifestyle for a fungus. How does it feed on spores? The hyphae of this strange fungus grow just beneath the underside of polypores and catch the numerous spores that rain down. *Sporophagomyces* pierces the spore cell walls and feeds on the contents.

ÖTZI, THE ICEMAN

In September 1991, hikers discovered the mummified body of a man emerging from a thawing glacier in the Tyrolean Mountains on the border between Italy and Austria. Dubbed Ötzi, the cadaver was initially thought to be that of a hiker who'd become lost and fallen into a crevasse. The body was transported to Bolzano, Italy, where researchers at the Archaeological Museum of Alto Adige made a shocking discovery—the remains were not of a recently deceased person but of a human male who lived between 3,300 and 3,100 BCE.

The physical and pathological examination placed his death, probably by violent means, between 40 and 50 years of age—quite old for that period. Based on pollen found in his lungs and stomach contents, the time of year of his death and his last meal, respectively, are known. We know much about his clothing, including the insulation stuffed into his footwear. He had tattoos. Ötzi was carrying a rich supply of artifacts: a bow, arrows, a piece of Birch Polypore (*Piptoporus betulinus*) used as a styptic, and inside a container was found a piece of *Fomes fomentarius*, wrapped in green leaves. The Tinder Polypore was no doubt smoldering inside at his time of death.

PUFFBALLS

Puffballs are fungal fruiting bodies familiar to most of us, since they often occur right out in the open on lawns and ball fields. They are saprobic and decay cellulosic debris like dead grass. Most are white and round—some can grow to enormous sizes. The most curious thing about them is: where do the spores come from?

AWAITING RAINDROPS

Puffballs have an advantage over many other types of mushroom-producing fungi, in that they shelter their hymenium within a chamber, protected from the drying conditions of the sunny open areas where they grow. When mature, the puffball skin (perideum) tears or breaks open to reveal the brown spore mass inside. Some small puffball species produce a small hole at the top for the spores to escape. When struck by raindrops, this mushroom—true to its name—puffs out clouds of spores that will land nearby and begin the next generation of its kind.

GIANTS OF THE FUNGAL WORLD

The surface of the Giant Puffball (*Calvatia gigantea*), also called *Langermannia gigantea*, is very round, smooth, and white when fresh, becoming olive to brown when mature. Average specimens are about the size of a football (12 in/30.5 cm in diameter), but can be basketball-sized or larger. This species is found in open woods or in fields or pastures in late summer and fall, and may even fruit in fairy rings. It is widespread and common across eastern North America and Europe. The Giant Puffball of western North America is *C. booniana*, which is noted for its more elongate shape and perideum that cracks and peels.

PUFFBALLS WITH PURPLE SPORES

The beautiful Purple-spored Puffball (*Calvatia cyathaformis*) can be quite small or nearly as large as the Giant Puffball at maturity. The base of this mushroom has a prominent lower portion that tapers gradually to a somewhat pointed base. It will become dark brown at maturity and, as its name indicates, has a spore mass that is purplish at maturity. *Calvatia craniiformis* is very similar but the interior never turns purple. Both are found across North America and reportedly southern Europe.

← When pelted by raindrops, Pear-shaped Puffballs (*Lycoperdon pyriforme*) emit puffs of spores that will germinate and begin the next generation of this mushroom.

STRANGE MUSHROOMS
(STINKHORNS)

All stinkhorns arise from a cuplike volva and produce a slimy, foul-smelling mass of spores on the cap. The mass of spores (called a gleba) attracts flies, which pick up the spores on their legs and bodies and carry them to distant locations. All stinkhorns are saprobic and rot debris found in parks and urban areas, farms and pastures, but can be found in woodland areas as well. After dining on the smelly gleba, the scavenging flies deposit it at the next rotting and putrid stop on their menu.

Many species of stinkhorns are placed in the genus *Phallus* due to their . . . obvious appearance. Mycologists are forever getting phone calls and emails about strange, phallic-shaped manifestations in their gardens or flower beds. And that smell! Shocking to some in appearance, they are, of course, totally harmless and, in fact, beneficial in breaking down debris and creating healthy soil.

← A freshly emerged stinkhorn mushroom like this *Phallus impudicus* will be covered with a goopy, muddy-looking mass of spores, known as a gleba, which will be quickly gobbled up by flies. Known as the Impudent Stinkhorn, its spores are dispersed by scavenging flies.

→ Stinkhorns are common urban mushrooms, living saprobically in organic debris. Some resemble undersea life, but many vary widely in look, from puritanical (complete with a golden or pure white veil like this beautiful *Dictyophora* species) to the prurient (unabashedly resembling sex organs).

PRIZED MUSHROOMS (TRUFFLES)

Truffles look like no other mushrooms. They more closely resemble small potatoes and indeed form entirely underground, termed "hypogeous." The famous and prized edible truffles of the world mostly grow around the Mediterranean and on the West Coast of continental North America, but many other regions, including China and the Middle East, have truffles that are commercially harvested.

TRUFFLING REQUIRES TEAMWORK

Truffles are mycorrhizal fungi, partnering with specific trees or other plants. Truffle hunters have long known this and, during the season, search around tree species known to host truffles. To home in on the exact location of a ripe truffle, hunters historically used pigs but nowadays mostly rely on dogs.

A LIFE UNDERGROUND

Truffles produce spores on a hymenium, like ascomycetes and basidiomycetes, but are underground. This poses a problem when the time comes to launch their spores into the air. Thus, they have evolved one additional trick: they produce strong savory odors that entice mammals to dig them up, consume them, and distribute their spores later on (with the rest of their solid waste). The different groups of fungi all feature species with traditional fruiting body forms. So why would some switch to a life underground? The vast majority of truffle species around the world are found in drier regions. Producing a fruiting body underground has the advantage of protection from drying out. The vast majority of truffle species are known from Australia, with more being discovered all the time.

FUNGAL TRUFFLE THIEVES

The largest and most commonly seen species of truffle parasites in North America and Europe seem to be growing from soil, but upon careful excavation, the collector will be delighted to find them attached to a deer truffle (*Elaphomyces* species). *Elaphocordyceps* (syn. *Tolycopladium*) *capitatum* has a discreet head atop the stalk where spores are produced; the fruiting body is directly attached to the deer truffle. In contrast *Elaphocordyceps* (syn. *Tolypocladium*) *ophioglossoides* (pictured) has an elongate head similar in appearance to the rest of the stalk; it is attached to a deer truffle by yellow hyphal cords (which may run several inches deep into the soil).

RUSTS, SMUTS, AND BLIGHTS

R ust and smuts are basidiomycetes—they produce spores from basidia—and are thus close relatives of mushrooms. However, they look nothing like mushroom-producing fungi. If you noticed them, you might wonder what on Earth they were—you may even suspect they were an alien life-form!

POORLY UNDERSTOOD

The rusts and smuts are very large and fascinating groups of fungi, but poorly known in many cases. All are parasites of plants and mostly pretty small. The rusts are especially interesting since many require two different plant hosts to complete their life cycle, while some have as many as five different spore types. Many rusts are important commercial pests of agricultural crops.

AROUND THE HOME AND GARDEN

Cedar Apple Rust (*Gymnosporangium juniperi-virginianae*) is a plant pathogen that produces weird, otherworldly life-forms on plants. This fungus, and other close relatives, are widespread across North America and Europe and may appear as jelly-like projections from the stem or branches of living tree host plants, or as ball-like galls with brightly colored jelly projections. Cedar Apple Rust may be seen wherever apples or crab apples and junipers coexist.

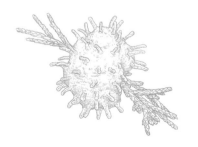

← If you notice galls with brightly colored jelly projections on your cedars, you may have Cedar Apple Rust. They occur wherever apples or crab apples and junipers coexist.

Corn Smut (*Ustilago maydis*) is a conspicuous parasite on living corn (maize) plants and may be seen throughout North America and Europe, but especially in the warmer regions. Historically, this fungus was common on field and sweet corn, but modern varieties have been bred for resistance. Heirloom corn is still very susceptible, however, as is popcorn and Indian corn. The fungus infection results in galls that develop on the ears, stalks, and tassels of the plant. The galls are edible and have long been considered a delicacy in Mexico, where they are known as *huitlacoche* (or *cuitlacoche*) and "Mexican corn truffle."

THERE'S A FAIRY RING ON MY ARM

Causing a sort of blight on our bodies, cutaneous fungi are often called dermatophytes for their propensity to live on skin. The afflictions they cause are known clinically as *tinea*, colloquially as "ringworm." There is no actual worm involved—think of it as a fairy ring on your skin. As the fungus grows outward, through the outermost layers of (mostly) dead skin, it causes a slight irritation that manifests as a reddened zone. The irritation in turn causes increased skin flaking, and thus more food for the fungus, and the dead skin that flakes away can help to spread the fungus to additional hosts. A common source of *tinea* transmission is via pets, especially dogs. The affliction can go by other names, according to where on the body it occurs—for example, *tinea pedis* for athlete's foot and *tinea capitis* for scalp infections.

CLAVICEPS PURPUREA: ERGOT

Claviceps purpurea is neither rust nor smut, but has a very similar life cycle. This ascomycete, known as ergot, produces hard, black bodies called sclerotia on rye or other cereal grains and grasses. These fruiting bodies are small and rarely noticed. Ergot is widespread across North America and Europe and is quite poisonous. One of the compounds in ergot is closely related to LSD and is responsible for hallucinations. Consumption of ergot sclerotia on contaminated grain and bread flour in the Middle Ages resulted in localized outbreaks of hallucinations, convulsions, and gangrene (see Chapter 11, page 141).

FUNGI EVERYWHERE

There is really nowhere on Earth you can go that fungi do not dominate or at the very least colonize. Wherever moist temperate climates favor life, the fungi are obvious: showy mushrooms erupt from soil and rotting wood. In the steamy, drippy tropics, fungi and lichens cover every surface and each other. Fungi are there in the drier and colder parts of the globe, too.

Terrestrial environments are much more the realm of fungi, of course, but there are also freshwater and even marine fungi. Fungi play crucial roles in nutrient and carbon cycling in aquatic habitats, just as they do in terrestrial ones. Some 3,000 species of aquatic fungi are recognized, of which only a small portion are known from marine waters. Given that 96.5 percent of our planet's water is oceanic, the species richness of marine fungi is likely to be much greater than currently known.

MICROHABITATS

Some habitats are likely to have niches or even microhabitats within them. These may be small natural features where abiotic factors such as light, temperature, moisture, and pH create growth conditions for unique and diverse life-forms not found in the larger environment. Microhabitats can involve animal hosts (birds and their nests, insects, and the rumens of herbivorous mammals); algae, lichens, and mosses; niches of rocks and caves; terrestrial, aquatic, and marine detritus; and so forth.

LOOKING FOR FUNGI WITH NEW TECHNIQUES

Soils in conducive habitats team with fungi. Indeed, fungi account for about half of all living organisms in soil. Although we are most familiar with macrofungi—those species that produce mushrooms and other fruiting bodies easily visible to the naked eye—they are but a fraction of all the fungi present. Most fungi are microscopic, most produce minute fruiting bodies, and still others seem to be entirely asexual, producing no fruiting bodies at all. We can see some of these hidden species back in the laboratory by culturing them from their substrates.

Many (maybe most) fungi cannot be cultured, however. Moreover, many habitats throughout all ecological zones remain unexplored. But ever newer techniques to discover fungi, including environmental DNA sequencing, allow scientists to "see" fungi in situations where they produce no visible fruiting bodies and cannot be cultured in the lab. Thus, the discovery of new fungal species is expected to increase.

ENDO- AND EPIPHYTIC FUNGI

Endophytic and epiphytic fungi—that is, the fungi that live within plants and on the surface of plants, respectively—have recently become a hot topic for research mycologists. Much remains unknown about these groups of fungi. Both fungal groups seem to play key symbiotic roles in the lives of their plant hosts, providing drought tolerance through plant-like hormones or by affording protection from mammalian and arthropod herbivory through the production of toxic compounds.

INSECT ASSOCIATES

Today, it is believed some 20,000 to 50,000 species of fungi associate with insects. Gut yeasts that help mycophagic insects digest chitin are believed to be a species-rich source of new fungi. Additionally, many fungi are known as insect parasites. Given that 80 percent of the estimated 5 million insect species remain undiscovered, it is likely these invertebrates will be another insect-related source of new fungi that could easily outnumber the entire insect population.

→ Wood rot fungi like the Honey Mushroom (*Armillaria ostoyae*) play crucial roles in nutrient and carbon cycling by breaking down dead organic matter in terrestrial habitats.

DESERTS

Fungi have adapted to life in extreme environments. In deserts, the fungi are there year in, year out, but may rarely emerge to form fruiting bodies. There are desert mushrooms and, as you would expect, if they show themselves at all, it's following an infrequent precipitation event.

As a result of evolutionary pressure, desert fungi produce sequestrate fruiting bodies—sort of puffball-like forms. Although many of these mushrooms are of the gilled sort, their caps never open (which would subject delicate hymenial surfaces to instant drying) and the gills never fully form. These fungi include *Battarrea*, *Podaxis*, and *Tulastoma* species, well known in Australia, North America, and Europe, wherever there is arid habitat. Still other desert fungi remain underground for their entire lives, even during fruiting. These include desert truffles like *Terfezia* and *Termania* species, among others. Popularly collected as delicacies, they fetch a high price in the markets and souks of the Middle East.

↓ *Terfezia* species are prized desert truffles that command a high price in the markets throughout the Middle East.

↓ *Helianthemum lippii*, a kind of Rockrose, is a host plant of desert truffles that pushes up and cracks the soil as it grows.

→ Desert fungi produce sequestrate fruiting bodies—puffball-like forms—that keep their spores sheltered within. When rains fall in desert habitats, *Podaxis pistillaris*, known as the Desert Shaggy Mane, emerge. As they mature, the cap breaks apart and sequestered spores are released.

ARCTIC AND EXTREME HABITATS

The last place on Earth you would look for fungi is Antarctica. Without question, this is the most difficult habitat for fungi, indeed all life, to survive—the Antarctic is literally the edge of life. But here, just as everywhere on the planet, fungi have figured out how to survive.

COPROPHILOUS FUNGI FIND A WAY

In the Antarctic there are few plants, thus not much in the way of sustenance for a saprobe to live on. But coprophilous ("dung-loving") fungi have adapted to life there. We mostly connect coprophilic fungi with the dung of grazing animals. In the case of marine birds in the Antarctic, their dung is almost entirely fish waste, and very high in proteins and amino acids, and some saprobic fungi there seem to thrive on it.

PARTNERING WITH AUTOTROPHS FOR SURVIVAL

Most Antarctic life is in the form of lichens, which are the dominant primary producers. Antarctic conditions are the most extreme in every way: perpetually cold, dry, and windy. Part of the year is completely dark; summers are light for nearly 24 hours a day and the UV is very intense, given the thin atmosphere and ozone layer. For life under these conditions, lichens have become endolithic. That is, they exist, amazingly, within the exposed porous rocks there.

Endolithic microbes exist about ½ in (10 mm) below the porous rock crust. The lichens are keystone species since they can photosynthesize. Microbial colonies can be distinguished by differently colored bands within the rock. A black band is melanized lichen and non-lichenized fungi; the melanin protects against the intense UV radiation. Below this layer you may find a green layer, comprised of non-lichenized photosynthetic algae and cyanobacteria.

As with Antarctic endolithic lichen fungi, melanins are key to survival. Melanins show high-tensile strength and thus reinforce cell wall structures such as spore walls. In this way, melanized fungal cell walls can better resist osmotic stress and turgor forces. Furthermore, melanized spores are much more resistant to desiccation and hazardous UV radiation. Hyaline, or colorless, spores are not typically viable for very long, but some fungi (for example, *Ganoderma*) produce darkly pigmented spores that can remain viable for years in soil.

Melanins even resist ionizing radiation, and are therefore important in our own epidermis. Remarkably, atomic radiation seems to cause enhanced growth in some melanized fungi, such as *Cladosporium* and *Penicillium* isolated from the Chernobyl reactor ruins. Amazingly, these fungi seem to harvest energy from ionizing radiation, making them autotrophic by a process that has yet to be understood.

↓ Plant and animal pathogenic fungi, like *Cladosporium uredinicola*, shown here, evade immune systems with their own defenses, including melanin, which gives the fungus a dark color.

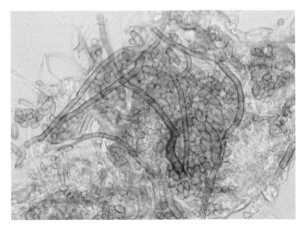

FUNGI IN URBAN AREAS

A number of fungi benefit from human practices like transportation, agriculture, and even landscaping. Indeed, many fungi are well suited to life decomposing the ubiquitous wood mulch so popular these days in the urban landscape.

Rarely noticed growing on wood chips and mulch are members of the genus *Sphaerobolus*. These tiny fungi are known as Artillery Fungi for their amazing ability to blast a spore packet (peridiole) great distances. True to their name, the force of the spore ejection even produces an audible sound!

Butt rots, fungal infections at the base of trees, are quite common in forests; probably even more so in urban areas where trees are damaged by vehicles or other machinery. A wound allows all manner of pathogens to enter, weakening the foundation of the tree and making it more susceptible to breakage and wind throw. Trees with these infections are hazards and their timely removal is necessary for public safely.

↓ In recent years, the Artillery Fungi have become a source of distress to homeowners, landscape mulch producers, and insurance companies. Due to the strong adhesion of the discharged peridioles—

they stick irreversibly to any smooth surface— they damage anything nearby on a sunny day. The pint-sized popguns have even marred the surfaces of vinyl siding on homes, windows, and automobiles.

→ Commonly seen in urban areas, butt rot fungi include some pretty large and sometimes showy polypores. These include species of *Inonotus*, *Laetiporus*, *Ganoderma* (pictured), and *Meripilus*, as well as *Onnia tomentosa*, *Heterobasidion annosum*, *Grifola frondosa*, and *Phaeolus schweinitzii*.

FUNGI IN OUR HOMES

O ur homes, gardens, and lawns are under attack. Despite our best efforts, more than 50 percent of the food produced on Earth is consumed by pests before it reaches our dining tables. Fungi need moisture to thrive, so preservation of food (as well as clothing, homes, and their contents) requires little more than maintaining dry conditions.

FUNGI CAN—AND WILL— CONSUME JUST ABOUT ANYTHING

If moisture is present, just about anything can be turned into a food source by fungi—that includes anything made of cellulose (cotton clothing, books, carpeting), wood, leather, or nearly any natural material. Priceless museum collections, antiques, libraries—virtually anything is at risk of damage. A number of fungi will grow on the materials of which the home is constructed, given the opportunity.

SAVING ANTIQUITIES

Some fungi, curiously, are found almost nowhere in nature but in human dwellings. This is especially troubling when destruction falls upon relics of historical or cultural significance. Mycologist Robert Blanchette is world-famous for researching and helping to preserve artifacts under attack from fungi. He is one of the few ever to be permitted to study the antiquities of the Forbidden City in China; he has studied fungal degradation in Sixth Dynasty Egyptian tombs dating from 4,000 years ago; and more recently he has been called in to investigate fungi at the South Pole. Amazingly, the fungi are not native there, but are of concern because they are causing destruction to wooden buildings abandoned during the "Golden Age of Discovery," when teams led by Robert Scott and Ernest Shackleton were in a race to discover the South Pole over land, a century ago.

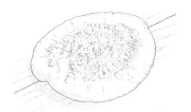

→ Pretty but so destructive, the Dry Rot fungus (*Serpula lacrymans*) almost seems to ooze over woody surfaces, where it weakens and ultimately destroys the integrity of the wood.

DRY ROT FUNGUS

Of all the wood decay fungi that cause damage to timber constructions worldwide, the Dry Rot fungus is considered the most destructive. This cosmopolitan fungus wreaks destruction the world over, causing billions of dollars in damage annually. The cause of this destruction, *Serpula lacrymans*, causes brown rot primarily of conifer wood. *Serpula* has the amazing ability to transport water (as well as nitrogen and other nutrients) by way of mycelial cords or rhizomorphs, often over great distances—even through the foundation of homes. Interestingly, the fungus can utilize quite a few inorganic materials for its nutritional needs, including calcium and iron ions extracted from plaster, brick, and stone.

FUNGI IN YOUR JEANS

Scientists have discovered ways to take advantage of some of these destructive fungi. *Trichoderma reesei* is used in industry to produce cellulase (enzymes that degrade cellulose). All strains of this fungus used industrially come from a single isolate collected in the Solomon Islands during the Second World War, where the fungus was the cause of a serious problem for the US Army. This fungus is especially good at making cellulase enzymes and destroyed the canvas tents used by the soldiers stationed in the damp jungles there. Today, the fungus is grown in huge tanks for the cellulases it excretes; much of the cellulase enzyme produced by this fungus goes to denim jean manufacturers. Those fashionable stonewashed jeans? Sometimes pumice is used to lightly abrade and soften the denim material, but much of the time the process involves cellulase enzymes instead to give the same results but at less cost to the manufacturer.

FIRE FUNGI

There is a group of fungi that are poorly known and rarely seen, for they show up almost exclusively following fire. Pyrophilous fungi have recently become the subject of much study, a result of consecutive years of unprecedented fires in many parts of the world, notably in Australia and North America. Where are these enigmatic fungi and what are they doing in the intervening years? And what is it about fires that promotes their growth?

Many of these fungi live as endophytes within lichens, trees, and other plants, in fire-prone areas. The majority of pyrophilous fungi are ascomycetes, as is the case with most of the lichen fungi. A few are basidiomycetes, including some species of *Pholiota*.

Possibly the most beautiful is *Geopyxis carbonaria*, commonly known as the Bonfire Cup and Pixie Cup. These stalked cups appear in spring right after a forest fire and are well known from all over the globe, from Australia to North America. And although the recently burned ground may be carpeted with this fungus, they pretty much only appear during that first year post-fire. After that they go back into hiding—they are not gone but going about their lives as an important symbiont of the forest. And waiting for the next big fire, their signal to spring into action.

STONEMAKER FUNGUS

One of the strangest mushrooms of Australia is also the most reclusive. In fact, the tuberous sclerotium of this fungus is more often encountered than actual fruiting bodies. *Laccocephalum mylittae* occurs in forests of south and eastern Australia. Early written accounts all state that Indigenous Australians regarded the excavated sclerotium as a delicacy, probably sliced and eaten raw, hence the common name for it: "native bread." Possibly the most interesting aspect of this mushroom is when it chooses to fruit. The sclerotia are thought to be perfectly happy growing underground for many years, maybe even decades, and appear to be an adaption to life in fire-prone habitats, for wildfire seems the catalyst for mushroom formation. Following the massive bushfires in 2019, mushrooms of *Laccocephalum* were commonly seen emerging from areas where the fungus was previously unknown.

← Bonfire Cups are often the first life to emerge from the ashes of wildfires. They are common all over the world, but often go unnoticed due to their small size.

FOSSIL RECORD

How old are the oldest fungi? How far back into the geological record do fungi go and how do we know this? Most people would assume that fungi do not fossilize. But it turns out that, although soft, fleshy fungi do not fossilize very well, we do have a fossil record for them. The first fungi undoubtedly originated in water, like much of the earliest life on Earth.

Based on the fossil record, fungi are presumed to have been present in the Late Proterozoic, 900–570 million years ago (Mya), and maybe further back than that. The oldest "fungus" microfossils were found in Victoria Island shale and date to around 850 million to 1.4 billion years old, although scientists are still debating whether these represent truly fungal forms.

MOLECULAR CLOCKS

Scientists use fossils to calculate very old dates. With fungi, most come from amber. Scientists also date organisms using "molecular clocks." This method has become indispensable for revealing the evolutionary pathway of the fungi. Molecular clocks are based on the rate of certain genetic mutations. All organisms are subject to alterations, or mutations, of their DNA. Mutations affect important and indispensable genes—changes to those affect the phenotype and are often lethal, so they alter very little through time. These are said to be highly conserved. Mutations also occur in neutral sequences, which do not affect the phenotype, and are therefore not under selective pressure. These sequences may change more frequently. And the knowledge that the mutation rate of such neutral sequences remains the same through time can be used to estimate the point when two species diverged in the course of evolution.

~ Fungi paved the way for plants ~

The first terrestrial plants date to around 700 Mya and the consensus seems to be that fungi probably arrived on land just ahead of them and paved the way for plants to move from marine to ever drier habitats.

A TALE TOLD IN AMBER

Much of what we know of no-longer-extant fungi comes from specimens found in amber. Amber is one medium that preserves delicate objects, such as fungal bodies, in exquisite detail. This is due to the preservative qualities of the resin when contact is made with entrapped plants and animals. Not only does the resin restrict air from reaching the fossils, it also withdraws moisture from the tissue, resulting in a process known as inert dehydration.

Furthermore, amber not only inhibits the growth of microbes that would decay organic matter, it also has properties that kills microbes. Antimicrobial compounds in the resin destroy microorganisms and "fix" the tissues, naturally embalming anything that becomes trapped there by a process of polymerization and cross-bonding of the resin molecules.

There are a few fossilized mushrooms known from specimens beautifully preserved in amber dating from the Cenozoic Era (starting 66 Mya) and Cretaceous Period (145–66 Mya). The oldest fossilized mushroom is called *Palaeoagaricites antiquus* (100 Mya) and resembles modern-day members of the family Tricholomataceae. Other mushroom species known from amber include *Archaeomarasmius legettii* (90 Mya), *Protomycena electra* (20 Mya), and *Coprinites dominicana* (20 Mya)—all look pretty much the same as mushrooms you would find in your local woods.

→ A rare find: a piece of 25-million-year-old Chiapas red amber, with a fossilized *Marasmius*-like mushroom inside. Only a few mushroom fossils within amber are known.

THE ARRIVAL OF FUNGAL GROUPS

The first "lichen-like" organisms we see date to around 600 Mya. About 550 Mya the chytrids and higher fungi split from a common ancestor. The first taxonomically identifiable fungi are from 460 Mya, and seem similar to modern Glomeromycota. Around 400 Mya the Basidiomycota and Ascomycota split from a common ancestor. The first insects came onto the scene around 400 Mya; the first beetles and flies date to around 245 Mya.

Mycorrhizas are extremely rare in the fossil record. Mycorrhizal relationships are believed to have arisen more than 400 Mya as plants began to colonize terrestrial habitats and are seen as a key innovation in the evolution of vascular plants. Recently, the first fossil ectomycorrhiza associated with flowering plants (angiosperms) was discovered. The fossils were found in a piece of Indian amber from the Lower Eocene (52 Mya), a time only 13 million years after the demise of the dinosaurs.

↓ Ascomycete fungi can form fruiting bodies such as these cups, *Microstoma protractum*. Their spores are then discharged from sac-like asci on the surface of the cup.

↓ Basidiomycete fungi form fruiting bodies like these Hedgehog Mushrooms (*Hydnum repandum*), with spores discharged from basidia on gills, tubes, or in this example, teeth.

→ Taxonomy separates fungi into the phyla Chytridiomycota, Glomeromycota, Basidiomycota, and Ascomycota based on their style of reproduction. Fungi have diverse forms (from the top): from primitive chytrids (e.g., *Chytriomyces hyalinus*) and zygote fungi (e.g., common bread molds like *Rhizopus nigricans*) to more modern mushrooms like the Chanterelle (*Cantharellus* species) and the Violet Coral (*Clavaria zollingeri*).

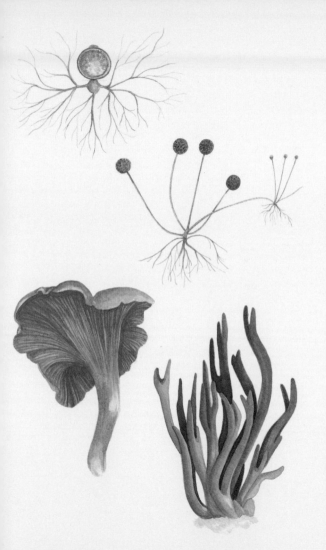

CONVERGENT EVOLUTION

Similarity of fruiting body forms can be misleading and has even led mycologists to disagree on classification schemes in the past. As a fascinating result of convergent evolution, ascomycete and basidiomycete fungi feature species that produce similar-looking mushrooms—for example, cups, clubs, and truffles.

Convergent evolution (sometimes called parallel evolution) is the evolution of similar traits in unrelated lineages. And it is very common in nature: animals, plants, and fungi all have many examples on display. What causes this? Convergent evolution is driven by the habitat and selective pressures on organisms—similar environments will select for similar traits in any species occupying the same ecological niche.

SIMILAR HABITATS RESULTS IN ANALOGOUS FORMS

Unrelated organisms often develop analogous structures by adapting to similar environments. Readily apparent analogous structures are seen in the animal kingdom: fins and wings. Wings can be found in unrelated groups of organisms like insects, bats, birds, and even extinct reptiles. Likewise, fins can be found in unrelated groups of organisms like fish, whales, penguins, and even extinct reptiles.

Convergent evolution works on the phylum level as well. Readily apparent analogous structures are seen among several orders within the phylum Basidiomycota. Many mushrooms produce shelf-like fruiting bodies on the sides of trees but not all are polypores. The environment and natural selection drives the organism into a form best suited for that niche.

NOT ALL CLUB FUNGI ARE ALIKE

The Purple Fairy Club (*Alloclavaria purpurea*) is a pretty mushroom and demonstrates convergent evolution. It appears to be one of the club fungi—the clavarias—common mushrooms in most temperate forests everywhere. Although it bears a striking resemblance to these fungi, it actually belongs to a larger group of fungi, the Hymenochaetes. Most members of this group are conks and brackets on wood that look and behave like polypores. But even they are masquerading . . . they are not true polypores. It is unknown what has driven these fungi to evolve their convergent forms. All are highly successful, based on their ubiquity in nature. And no one can deny their beauty.

TRUFFLES AND FALSE TRUFFLES

Truffles are another example of convergent evolution, where organisms from quite different evolutionary lines have independently discovered a similar way to make a living. "True" truffles are ascomycetes; they produce spores in elongated sacs and include the morels and cup fungi. "False" truffle is a term reserved for truffles that are not ascomycetes.

Many groups of basidiomycete fungi (mushrooms with a cap and stem) have lost their "normal" shape and gone underground (become hypogeous). Both feature a layered or convoluted hymenium (where the asci and basidia produce asco- and basidiospores, respectively) that is sequestered inside and not exposed. Clearly the truffle form is adaptive, but how; and how did it come about?

FORM FOLLOWS FUNCTION:
THE CONVOLUTED HYMENIUM

Although not obvious, true truffles (*Tuber* species) are most closely related to members of the order Pezizales, which includes *Peziza*— very large, brown cups, commonly seen on woody debris. How did one branch of ascomycete fungi go from a flattened morphology and epigeous (above ground) growth habit to highly convoluted and hypogeous (subterranean)? Taking the reproductive surface layer, or hymenium, and convoluting it allows for more surface area (and more spore production) per unit area of mushroom. Morels are another group of ascomycete fungi that do this (but they are epigeous).

Truffle-like forms have evolved several different times within the Basidiomycota. In fact, for just about any common genus of mushrooms, we could follow an evolutionary progression from "typical" mushroom morphology to ever more truffle-like. As one example, milk mushrooms (*Lactarius* species) have a sequestrate form, *Arcangeliella*, and a very truffle-like form, *Zelleromyces*; and just like their milk mushroom relatives, both exude a milky latex. In fact, no fewer than 14 families of mushrooms have separately given rise to sequestrate or false truffle forms.

SUBTERRANEAN LIFE

There are several possible reasons why truffle forms adopted an hypogeous existence. Various groups of hypogeous fungi may have been driven underground by some biotic factor like mycophagy; maybe mushroom-grazing animals were consuming too many fruiting bodies for that style of reproduction to be successful within the group. More likely it was due to environmental, or abiotic, factors. Most fungi are very sensitive to dry conditions, especially when forming the fruiting body. It is probable that development of a sequestrate and hypogeous fruiting body gives those fungi an advantage in drier areas. Drier habitats of Earth seem to favor truffle forms; Australia is thought to have far and away the most species of truffles and false truffles.

~ Spore dispersal ~

Producing spores in a subterranean fruiting body presents new challenges: namely, how to disperse those spores. Truffles entice other animals to help. Several mammals like deer and squirrels dig up and consume truffles. Many invertebrates are truffle feeders, including slugs and insects; many fly species are probably strict truffle feeders.

TRUFFLES DRIVE MAMMALS WILD

Tuber species attract mammalian vectors by producing a smelly compound called alpha-androstenol. This chemical is also found in the saliva of rutting boars and acts as a pheromone to attract sows. Many other mammals probably also produce this pheromone, which explains why numerous digging mammals are attracted to these fungi.

↙ Truffles are simply underground mushrooms. The spores are produced on the convoluted surface within.

CO-EVOLUTION AND MIGRATION

During his voyage on HMS *Beagle* in 1839, Charles Darwin (1809–1882) collected a peculiar fungus from large cankers on trees during a stop at the southern tip of South America. Resembling alien life-forms, those brightly colored *Cyttaria* fruiting bodies are relatives of morels. Indeed, both are apothecia, a sort of cup-shaped ascocarp, with sterile ridges separating the fertile areas.

Cyttaria are obligate biotrophs of Southern Beech (*Nothofagus*) and are restricted to the Southern Hemisphere, inhabiting southern South America and southeastern Australasia. The relationship of this fungus with its host tree remains unclear. And it's hardly the only strange aspect of this fungus.

How has *Cyttaria* spread all across the vast oceans of the Southern Hemisphere? *Cyttaria* had co-evolved—and been geographically isolated on landmasses—with their respective host species of *Nothofagus*. Thus, species of *Cyttaria* and *Nothofagus* have not actually moved anywhere at all—they've been stuck with each other since the breakup of Gondwanaland, more than 200 million years ago.

↓ *Nothofagus* trees have been well-studied (more so than their fungal partners). Their geographic range is directly linked to the movement of Earth's landmasses. This time-lapse series shows the movement of landmasses over a period of 200 million years, until the final image of modern Earth.

→ Young fruiting bodies of *Cyttaria darwinii* are smooth and firm, but later develop numerous fertile pits once the membrane bursts. These ascomycete mushrooms produce spores like morels and other cup fungi.

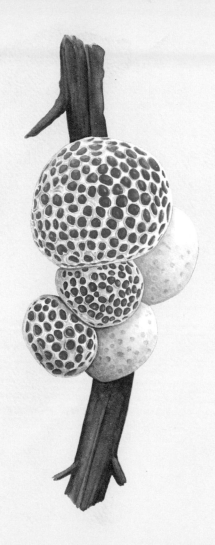

ONE PATHOGEN, DIFFERENT HOSTS

R ust fungi (order Pucciniales) are fascinating and numerous (about 7,000 species in 168 genera worldwide). And enigmatic: rusts have up to five spore stages (spermagonia, aecia, uredinia, telia, and basidia in successive stages of reproduction). They are all obligate parasites, requiring a living host. Most rust fungi that infect trees have spore stages on two completely unrelated hosts.

RUST OF LUMBER TREES

White Pine Blister Rust (*Cronartium ribicola*) is among the most famous of forest tree diseases. Native to Asia and introduced to North America in the early 20th century, it arrived on pine seedlings from France. The fungus causes elongate perennial cankers on stems and branches; its complex life cycle requires two hosts: a white pine and a bush (*Ribes* species, currant or gooseberry).

This disease is very important economically, and to stop outbreaks the disease cycle must be broken. The US government launched a program to eradicate wild currant and gooseberry plants in the East, lasting from the 1920s to the 1950s, significantly reducing the population of *Ribes*. The federal ban on *Ribes* cultivation and sale was lifted in the 1960s, but state quarantine laws still exist today in many eastern states, where white pine is an important plantation tree.

→ Wheat Stem Rust (*Puccinia graminis*) requires two very different host plants to complete its life cycle. Light microscopy reveals spore production from wheat tissues.

RUST OF CEREAL GRAINS

Rusts are responsible for the most important diseases of many of our cereal grains, including Wheat Stem Rust and Crown Rust of oats. Rusts of wheat plants, caused by the fungus *Puccinia graminis*, could be the most economically important disease of plants, globally. This disease has annually caused losses in excess of 1 million tons in North America, but in severe epidemic years losses reach tens or hundreds of millions of tons. As the world becomes more crowded, and hungry, this fungus will certainly be the cause of mass famines and even wars.

ROBIGALIAS

Human civilizations have struggled with Wheat Stem Rust for centuries; the Romans tried to appease the fungal gods with "robigalias"—elaborate ceremonies in which dogs were sacrificed in an effort to stave off the rust-colored "red fire" that annually descended upon and consumed their fields of wheat.

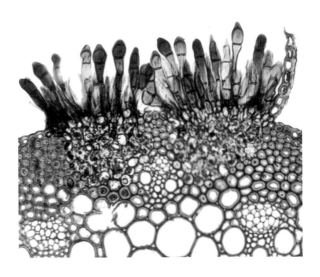

OUR ROTTEN WORLD

Next time you encounter a fallen tree in the forest, examine it. You will find a tiny ecosystem. Ultimately, sunlight was the original input of energy into that system; the tree captured it autotrophically. Now in death, all that organic matter—likely tons of carbohydrates and proteins and other building blocks of life—are sitting there for the taking, for any organism with the ability to break down wood. Simple single-celled bacteria can ingest sugars sitting on the surfaces of the wood; slime molds ooze over and engulf them.

Fungi are well adapted to break down wood using cellulase enzymes; wood-boring beetles, wood wasps, and other arthropods can feed on degraded wood inhabited by fungi—much of the time, the fungi were inoculated into the wood by their insect partners. Woodpeckers, other birds, and mammals tear through the wood in search of arthropods to dine on. Other members of the forest make homes out of cavities in the wood. The circle of life for the ecosystem that was a single tree is complete when the log serves as a nurse tree for seedlings or is decomposed entirely back into soil.

↙ Fungi are good at decomposing wood, but it can take a long time. The Dyer's Polypore (*Phaeolus schweinitzii*) is commonly found on conifer wood.

WOOD DECOMPOSITION

Decomposition of carbohydrates and other organic matter is pretty much the same chemical process as photosynthesis, but in reverse. During photosynthesis, plant chlorophyll captures the energy of red and blue wavelengths of sunlight (green is not much used and is reflected, with the result that plants appear green), and converts that into energy to "fix" single carbon molecules (the very plentiful carbon dioxide of the atmosphere) into growing chains of carbons and hydrogens—literally the carbohydrates of plant matter. One by one, carbons are fixed into six-carbon sugars; these are linked together to form cellulose and other carbohydrates, and plants grow.

Plants are mostly composed of cellulose and water. Woody plants or those with a rigid structure also contain a great deal of lignin. Lignin is a polymer of very tough ring molecules crosslinked in a random fashion that strengthens wood. Both cellulose and lignin are difficult to break down and require arsenals of enzymes and other machinery. For the most part, wood-degrading fungi are good at breaking down one or the other, but they're largely after the same thing: cellulose.

FOSSIL FUELS: A THING OF THE PAST

We are truly fortunate that fungi came along. Without them, we would surely be buried under miles of dead organic matter. Fungi do the planet a great service by breaking down and recycling the dead organic matter that accumulates all over Earth, most of which is cellulose and lignin from dead trees and other plants. That we can dig up vast pockets of buried coal and petroleum is evidence that, at one time, there were no wood rot fungi.

Scientists have determined that fungi capable of rotting woody plants didn't come onto the scene until the end of the Carboniferous Period (360–290 Mya)—quite a bit after the evolution of woody plants. So instead of decomposing, all that organic matter piled up and changed by a process of chemical reduction (the opposite of oxidation), and thus fossilized—into fossil fuels like coal. With the proliferation of wood rot fungi, the buildup of coal deposits dramatically decreased during the Permian Period (299–251 Mya).

BROWN ROT AND WHITE ROT

Plants are mostly made up of cellulose and lignin. Both compounds are made of carbohydrates but are very tough to break down. Armed with wood-degrading enzymes, fungi are just about the only organisms that can do this.

BROWN ROT FUNGI

Fungi that directly break down cellulose, leaving the brown lignin behind, are called "brown rot" fungi. The removal of the cellulose destroys the structural integrity of the wood and it cracks and falls apart as cubes. Examples found worldwide are the polypores *Laetiporus*, *Phaeolus schweinitzii*, and *Fomitopsis*.

WHITE ROT FUNGI

In contrast, the "white rot" fungi decompose lignin, bleaching the wood and initially leaving stringy, white cellulose behind. These fungi have powerful peroxidase and laccase enzymes that break down the lignin. While there is evidence that they can decompose lignin completely to carbon dioxide, many researchers suggest the fungi are mostly removing it from the woody pulp to better get at the cellulose. White rot fungi found worldwide include polypores like *Inonotus*, *Ganoderma*, and *Trametes*, as well as *Pleurotus* and *Armillaria*. The popular cultivated Shiitake mushroom (*Lentinula edodes*) is also a white rot fungus. The power to bleach wood pulp of lignin makes the white rot fungus *Phanerochaete chrysosporium* important to the paper industry as an environmentally benign replacement for harsh synthetic chemicals.

→ The white, stringy debris resulting from the peroxidase and laccase enzymatic action of white rot fungi. Decomposition of lignin leaves the white cellulose behind.

WOOD ROT ATTACKS LIVING TREES

Many of the wood rot fungi familiar to us (for example, the big polypores) don't wait for trees to die before launching their assault. Living trees often sport big fruiting bodies of shelf fungi. This is because most of the tree is heartwood—the dead inner wood. (Only the outer layers, just under the bark, are living tissue.) All it takes is a wound to disrupt the integrity of the bark and heart rot can ensue. Similarly, butt rot happens at the base of the tree.

Heart rot can proceed for many years without really having much negative effect on the tree. Upon seeing large polypore shelves hanging off the side of a tree, most people assume the fungus is a parasite of the tree. In reality, few polypores are truly parasites of living tissue. An otherwise healthy tree takes measures to contain these heart rot fungi, preventing them from invading living tissues in most cases.

FUNGI WITH A UNIQUE LIFESTYLE

N o matter the source of nutrition, there is a group of microbes to utilize it. Some components of animals persist long after death, including those constructed of keratin like fur, feathers, and horns. Keratinized material is so tough that only a single group of fungi can decompose it: the order Onygenales.

Undoubtedly the most unusual but least understood genus of the group is *Onygena*. If on the odd chance you happen to find an animal's horns lying in the forest, examine them. They may be covered with tiny, stalked, mushroom-like fruiting bodies, a sure sign of *Onygena*. (Antlers are bones and not made of keratin.) Owls and other birds will regurgitate a pellet of bones, fur, feathers, toenails, and other undigestible material. These are good places to find *Onygena*, as well as to see what's on the menu of predators in your area.

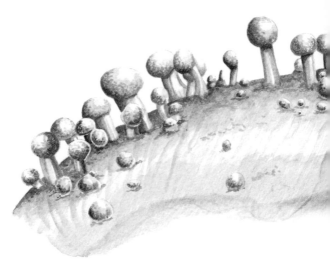

⬐ The genus *Onygena* consists of just two species: *Onygena corvina* is associated with animal feathers and fur, while *O. equina* is a decomposer of the hooves and horns of herbivorous mammals.

⬐ Close examination of *Onygena equina* sporophores will reveal that what appear to be tiny, stalked mushrooms are, in fact, masses of spores at the tips of aggregations of hyphae.

ANIMAL PARASITES

Fungi are now routinely making the headlines and the news is not often good. We are currently facing two major animal crises: a massive decline in amphibian species and an explosive disease outbreak among bats in North America. What a terrible loss to the future planet it would be if there were no amphibians to sing in our wetlands and no bats in our evening skies.

GLOBAL AMPHIBIAN DIE-OFFS

For many years, herpetologists around the world had noticed that amphibian populations were in decline, but the evidence remained largely anecdotal. In the 1990s, an unknown fungal disease was identified—chytridiomycosis—which was causing widespread amphibian mortality in Australia, as well as North, South, and Central America.

The causative agent is *Batrachochytrium dendrobatidis* (Bd). The lifecycle of Bd involves a motile, swimming spore that finds a host animal and sticks to its skin. In a matter of days, new zoospores are released, which swim about and further infect the same host; if they find another amphibian, a new infection begins. When covered in fungal zoospores, the host is unable to breathe and dies. Bd is responsible for what is perhaps the largest panzootic in history and infects many frog, salamander, newt, and other amphibian species. Undoubtedly, some amphibians will be wiped out, but immunity is cropping up in various places and some amphibians are beginning to bounce back.

← In the United Kingdom, most populations of the native Natterjack Toad (*Epidalea calamita*) have tested positive for the fungal disease chytridiomycosis, but are apparently unaffected.

WHERE DO THESE PANDEMICS COME FROM?

Both pandemics are not fully understood. Chytrid fungi may have long been associated with amphibians. Increasing UV sunlight levels and global climate change may stress amphibians, allowing these fungi to become pathogenic. However, *Batrachochytrium dendrobatidis* (Bd) could be new: examination of preserved amphibians in museum collections found no Bd prior to 1938, which corresponds to the inception of trade in African Clawed Frogs (*Xenopus laevis*) used in research labs and pet aquaria.

White-nose Syndrome (WNS) seems to be a newly emerging disease from Europe—it is found in caves throughout Europe, although doesn't seem to cause problems for bats there. It's likely that European bats have been around the fungus for millions of years and have had time to develop resistance to it.

EPIDEMIC BAT DIE-OFFS

In late winter 2007, researchers found thousands of dead Little Brown Bats (*Myotis lucifugus*) with a white growth on their muzzles and ears in five caves in upstate New York. Bat White-nose Syndrome (WNS), caused by *Pseudogymnoascus destructans* (Pd), has since spread throughout eastern North America. The fungal pathogen is known to infect at least 13 species and millions of bats have been killed. Since bats pollinate some plants and eat pest insects, their value to US agriculture has been estimated at least $3.7 billion a year.

Pseudogymnoascus destructans is a saprobe living on organic matter found in caves. Its growth on living bats is still somewhat of a mystery, and seems to be opportunistic. Growth on the skin of bats seems to irritate them out of hibernation, causing them to fly about. This activity consumes winter reserves bats can ill afford. If they leave the cave before spring, they waste further energy in a vain search for food. Thus, bats that succumb to WNS often die of starvation.

COPROPHILOUS FUNGI

Spores of coprophilous (dung-loving) fungi are tough and resistant, with thick walls adapted to survive passage through the intestinal tracts of herbivores. This makes them conducive to fossilization and they are often found deep in soil layers, lake deposits, and permafrost. Scientists use these fungi as a tool to study mammals long gone.

Spores of *Sporormiella* are unmistakable in soil sediments—even from thousands of years ago. Archaeologically, these spores are an indicator of vegetation changes throughout history. Accumulations of *Sporormiella* spores are correlated to an abundance or absence of herbivores today, as well as in the past.

Scientists can determine when the mammalian megafauna dominated North America, and when they began their decline due to factors like a changing climate and Paleo-Indian hunting pressures at the end of the Pleistocene Epoch (2.6 million to 11,700 years ago). Following the last Ice Age, the numbers of these dung fungi remained low until 17th-century European settlers brought livestock to North America.

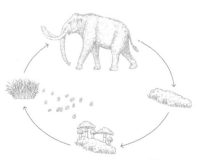

↓ The spores of dung fungi are inadvertently consumed along with grass and pass through the herbivore's gut; the spores are then discharged with feces. These germinate in the new substrate and the life cycle continues.

→ The undigested plant matter that passes through grazing animals is mostly cellulose, and therefore it is highly nutritional for dung-loving fungi, including *Deconica coprophila*, shown here. Since the animal did much of the mechanical work by grinding and partially breaking down this material, being the first to colonize dung before competition arrives has led to some interesting specialization, e.g. *Pilobolus* lifecycle.

BIOLUMINESCENCE

Glowing mushrooms are amazing to see ... and mysterious. Bioluminescence has been documented since the time of Aristotle (384–322 BCE) and Pliny the Elder (23–79 CE). The four lineages of bioluminescent fungi contain around 80 different species. Mushrooms familiar to us that glow include *Armillaria*, *Mycena*, *Omphalotus*, and *Panellus*. "Foxfire" is glowing deadwood, a sure sign of wood decay by hyphae of *Armillaria*.

WIDESPREAD IN NATURE

Bioluminescence is common in the natural world; in addition to fungi, some animals, plants, and bacteria can do it. Two things to keep in mind about bioluminescence are that it is ongoing, even in the light of day, although not visible. Likewise, bioluminescence generates no heat and is thus very different from incandescence, which is a thermal glow. Today it is known that the light originates from a metabolic reaction of the fungus where electrons are transferred to an acceptor molecule (luciferin), which is cleaved by an enzyme (luciferase) in the presence of oxygen. This results in the formation of an electronically excited state of the luciferin and the subsequent emission of light with a maximum wavelength of approximately 525 nanometers (nm) during return to the ground state. This process is much the same for all organisms that bioluminesce, although the luciferins and luciferases are not exactly the same.

→ *Omphalotus* species are wood rot fungi that seem to produce large, but pretty drab mushrooms when seen in the light of day. But after dark, their eerie green glow can be properly seen.

FUNGAL BIOLUMINESCENCE
REMAINS AN ENIGMA

What is the purpose of bioluminescence? Might it serve as an attractor of invertebrates for the purposes of spore dispersal? Although the subject of much research over many years, we still don't know. Bioluminescence may simply be a way for fungi to dissipate energy as a by-product of oxidative metabolism (most organisms, including us, give off heat as a by-product). Moreover, this chemical reaction may be tied to detoxification of the peroxides formed during lignolysis. Many bioluminescent fungi rot wood and leaf litter—for example, *Armillaria mellea* and *Panellus stipticus* are white rot fungi. There are genera with glowing and non-glowing species that all seem to be equally successful in nature; thus, many argue it serves no selective advantage. Within the genus *Mycena* there are at least 33 species, from 16 sections, known to bioluminesce.

THE SEX LIVES OF FUNGI

Fungi reproduce very differently from other life-forms. Most, if not all, fungi can reproduce asexually. A benefit of sexual reproduction comes from genetic recombination (which is undoubtedly why it is so common among eukaryotes). Also termed "crossing over," offspring receive a combination of DNA from both parents of that individual. Sexual reproduction in all organisms involves three events: the fusion of two haploid (n) cells (plasmogamy); fusion of nuclei (karyogamy) to form a diploid ($2n$); and meiosis that results in recombinant haploid nuclei that will be offspring.

Animal cells are diploid dominant, and during sexual reproduction haploid gametes of two parents fuse to restore the diploid state. Most fungi exhibit a life cycle with both a haploid and a diploid phase; for many, each cell contains two haploid nuclei (one from each parent) as a dikaryon.

BASIDIOMYCETES DO IT

Basidiomycetes can live as a dikaryon for a long time; only at the time of sexual reproduction does karyogamy take place (in the basidium) of the fruiting body and results in haploid spores (basidiospores). These germinate and fuse with an opposite mating type hypha pretty quickly in order to restore the dikaryon state and a full complement of necessary genes.

ASCOMYCETES DO IT

With ascomycetes, the timing of these steps depends on the group. Many common ascomycetes are anamorphs (asexual forms), and seem to do just fine as monokaryons. In teleomorphic (sexual) ascomycetes, the fungus can undergo plasmogamy and live as a dikaryon. At the time of sexual reproduction, karyogamy takes place in the ascus of the fruiting body. Karyogamy and the diploid state is quickly followed by meiosis and haploid spore production.

TWO GENDERS? . . . FOUR? . . . 28,000?

Some fungi are self-fertile (homothallic), while many are heterothallic and must out-cross. For the latter, sexual reproduction is governed by compatibility or mating type genes. Mating type genes allow a fungus to mate with all members of the same species except those of the same mating type—essentially a mechanism to avoid inbreeding. Many fungi have just two or four mating types, and so on; at the extreme end, *Schizophyllum* has 28,000 mating types. This amazing discovery, along with the fact that there is but a single species of *Schizopyllum commune* (shown above) found worldwide, was made in the lab of John and Carlene ("Cardy") Raper, both of whom devoted their lives to the study of this basidiomycete fungus.

SPORE PRODUCTION

Historically, spore release was assumed to be passive: spores drop from the mushroom and waft away on air currents. Not so! Basidiomycete spore release is explosive. Called ballistospory, it's more like a surface tension catapult. The spores (ballistospores) are borne on the surface of gills or tubes of the mushroom. Lining the hymenial surface are specialized hyphal tips called basidia. Atop each club-shaped basidium are outgrowths known as sterigmata where spores will develop.

BULLER'S DROP

Key to spore ejection is the production of Buller's drop. At the sterigma, hygroscopic liquid is released and moisture from the air condenses on this liquid. The condensate grows into a droplet—Buller's drop—until it reaches a critical size, at which point it touches the water film on the spore surface and coalesces. Surface tension quickly pulls the drop onto the spore and kinetic energy creates the necessary momentum to detach the spore from the hymenial surface. Ballistospores are blasted from their basidia, then fall vertically under the action of gravity and are carried away by air currents.

Named for the British-Canadian mycologist Reginald Buller (1874–1944), Buller's drop formation requires moisture from the air, but too much water can disrupt this mechanism altogether. For this reason, many basidiomycetes have fruiting bodies that are umbrella-shaped to shield the hymenium from rain. Many others sequester the hymenium altogether within a fruiting body, like a puffball or a truffle.

→ Although they are microscopic, copious amounts of mushroom spores can pile up beneath a cut mushroom cap left on a surface overnight, resulting in a spore print.

THE FASTEST FLIGHTS IN NATURE

Since spores are discharged at extremely high speeds, researchers could only calculate launch speeds using mathematical models. (Many probably just guessed.) In 2008, mycologist Nicholas Money and colleagues used ultra-high-speed video cameras running at rates of 250,000 frames per second to analyze the entire launch process in four species of fungi that grow on the dung of herbivores. For the first time ever, they were able to take direct measurements of launch speeds of fungal spores. How fast? Dubbed "the fastest flights in nature," launch speeds ranged from 2 to 25 meters per second with incredible corresponding accelerations of 20,000 to 180,000 times the force of gravity!

PASSIVE SPORE DISPERSAL

Bird's nest fungi are very commonly found on woody mulch in urban settings. They produce their spores in small packets (peridioles) that appear as "eggs" in a nest-like cup. A raindrop hitting the cup will splash the eggs great distances away.

Bird's nest fungi are tiny (less than ½ in/12.5 mm) and are widespread across North America and Europe. The Common Bird's Nest (*Crucibulum laeve*) is yellow and cone-shaped, while the Fluted Bird's Nest (*Cyathus striatus*) is dark with a grooved or striate cup interior.

In nature, these fungi are very successful at rotting small sticks and twigs in the canopy overhead (although you are more likely to see them after this debris falls to Earth). With all that splashing, how do they stay up in the canopy? Peridioles splashing free from a cup unfurl a long elastic cord of hyphae known as a funiculus. This wraps around the first twig it comes to, much like an anchor.

↓ Bird's nest fungi are decomposers that produce spores in egg-like packets (left) that splash from cups. Upon ejection, a tiny

anchor trails behind (center), which snags on plant matter nearby (right). The fungal spores germinate and begin life for the next generation.

→ Looking every bit like eggs in a real bird's nest, tiny fungal peridioles of *Crucibulum laeve* await a well-placed raindrop to be ejected from mature splash cups; immature cups are covered.

ASEXUAL OOMYCETES

There have been episodes in our past where millions of human lives have been lost because fungi and fungi-like pathogens have wiped out crop plants, causing mass starvation. Probably the most infamous has become known as the Great Famine, or Irish Potato Famine, which hit Europe in the mid-1800s.

In Ireland the impact of the Great Famine is still noticed today; over just a few years 1 million people starved to death and another 2 million (or more) emigrated from Ireland. The population of Ireland has never totally rebounded and is today still far lower than it was prior to the potato famine. Before the famine, Ireland's population was around 8 million, while today it's around 6 million.

Late Blight disease is caused by the oomycete *Phytophthora infestans*, and it can sweep through an entire crop with amazing speed. All it takes is a single spore or hyphae remaining in plant residue or a single tiny tuber left over from the previous crop. In as little as one week, the pathogen can totally destroy an entire field if conditions are cool and wet. Asexual reproduction affords swift spore production—and without the need nor time to find opposite mating types.

← Disease cycle of Late Blight infection spreads quickly via motile zoospores. All parts of the potato plant may be infected, including the leaves, stems, and tubers.

GENETIC ISOLATION BREAKS DOWN

Following the worst years of Late Blight in the 1800s, scientists began to get the upper hand with the development of fungicides as well as through classic plant breeding, which had produced several potato cultivars resistant to Late Blight. That all came to an end in the 1980s when, out of nowhere, the pathogen swiftly broke down resistance to fungicides and broke through resistant potato varieties. What had changed? *Phytophthora infestans* is capable of sexual reproduction but until the 1980s this had rarely been seen and was hardly known for centuries. During trade with Europe, the Late Blight pathogen was introduced, but only one of the two mating types, named A1. The A1 strain of *P. infestans* was active there for decades; destructive to be sure but reproducing asexually only. A second mating type, A2, finally found its way to Europe in the 1980s and to North America soon after. With sexual reproduction came genetic recombination and today we are once again faced with the very real possibility of complete destruction of potato crops.

NEWLY EMERGING EPIDEMICS

Newly emerging tree diseases are causing alarm and have researchers on their heels. Sudden Oak Death (SOD) causes a lethal infection of the trunks of several species of oak and has killed hundreds of thousands of trees in the few years since it turned up in California in 1995. The pathogen also sickens several other unrelated species, including azaleas, rhododendrons, *Viburnum*, larch, and maples. The cause of SOD is the oomycete *Phytophthora ramorum*; the pathogen is a problem in Europe as well, and it's possible it came to North America from Europe. Despite a quarantine in 2001, SOD has spread up the West Coast and moved into British Columbia. States across the United States have imposed bans on all nursery stock from California, but every two or three years infected material escapes quarantine. Many fear the pathogen could spread to forests of the Southeast and elsewhere, and cause untold destruction.

ACTIVE DISPERSAL AND ZOOCHORY

The two major modes of fungal spore dispersal are wind- and animal-mediated dispersal (zoochory). Zoochory is much less common; of the tens of thousands of species of fungi known, only about 1 percent have an association with animals. For fungi producing sequestrate and hypogeous sporocarps, like truffles, there must almost always be an animal vector to assist with spore dispersal. Ingestion and subsequent defecation is known as endozoochory.

Animals, including insects, haphazardly transport spores and yeasts. Many fungi, however, have evolved elaborate strategies to entice animal vectors. Some groups of insects are obligately associated with fungi as a food source and ensure that the fungus goes wherever they go. And so they have evolved pouches, called mycangia, on their bodies specifically for this purpose.

Xylophagous (wood-boring) insects cannot actually digest woody cellulose but rely on fungi—or their enzymes—to break down the wood for them. Some insects inoculate wood and, after a period of time, will begin feeding on the now-digestible wood. Some of these fungal symbionts are plant pathogens that attack and weaken the host tree, making it more prone to beetle attack.

AMBROSIA BEETLES

The most fascinating of xylophagous insects are the ambrosia beetles. They bore through wood and inoculate it, then spend their lives feeding exclusively on the fungal gardens growing on the walls of their sapwood galleries and tunnels. Ambrosia fungi are adapted to symbiosis with the beetles. These fungi are dimorphic, growing as filamentous hyphae within beetle galleries where they produce a dense layer of easily grazed conidiophores ("ambrosia") that the beetles feed on. Mycangial transport of these species is thought to be the sole means of transmission for these fungi, as they live trapped deep inside beetle tunnels. The second of the two life-forms of ambrosia fungi is a yeast-like morphology that is cultivated and nourished inside the mycangium by glandular secretions of the beetle.

WOOD WASPS

Bark beetles, termites, and ants are not the only insects to have co-evolved with fungi. The Giant Horntail wasp (*Tremex columba*) is a very large (about 2 in/5 cm long) wood-boring wasp (family Siricidae). There are countless species of wood-boring insects on the planet and in all cases, including siricid wasps, none of them make enzymes to digest wood cellulose. Instead, siricids rely on basidiomycete white rot fungi, *Cerrena unicolor*, even transporting these fungi to the wood source. These symbioses are mutualistic, as both partners benefit: the wasps are able to utilize a large energy resource in the forest and the fungus benefits not only by being transported to a specific host tree, but also by the insect carrying the fungus past the tree's first line of defense (the corky bark) and into the interior wood.

In the case of the horntail wasp, conidia and mycelium are carried within the female wasp's mycangium. When the horntail lays eggs inside the tree, conidia pass from the mycangium through ducts to the ovipositor. No doubt *Cerrena* fruiting bodies produce heavy loads of basidiospores that result in successful infestation of wood, but compared with the insect as vector, the former seems much more haphazard. And without the involvement of *Tremex* species, infestation into a living tree is probably even less likely.

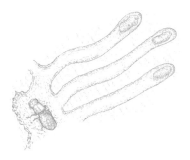

← Adult bark beetles carry wood rot fungi to dead wood and lay eggs. Their larvae tunnel through and consume wood that has been partially broken down by the action of the fungi.

ACTIVE DISPERSAL AND MIMICRY

In New Zealand where birds dominate food webs, sequestrate and truffle fungi produce fruiting bodies (sporocarps) that resemble berries on the forest floor. Foraging birds gulp them down and deposit their spores elsewhere. Possibly the most convincing "berry truffle" is the Scarlet Berry Truffle (*Paurocotylis pila*). As its hypogeous fruiting bodies mature, they emerge from soil and resemble the fallen red fruit of *Podocarpus* trees that mature and drop at the same time.

Paurocotylis is a strange fungus that relies on zoochory and mimicry during reproduction. In Oceania, the truffle-form is a much more common fruiting body morphology than anywhere else. Digging mammals are the most important vector of spores for truffle-producing fungi of Australia. Truffles that rely on mammals for dispersal are mostly dull colored but produce very strong odors, as mammals largely forage using their sense of smell. Conversely with birds, brightly colored purple, blue, or red fruiting bodies that resemble berries lying on the forest floor are what catches their eye.

↓ Everything about the Scarlet Berry Truffle is strange, from its poorly known life cycle to its truffle-like fruiting body. The fruiting bodies of this fungus appear to be brightly colored berries, but a cross-section reveals chambers lined with spore-producing hymenia.

→ Scarlet Berry Truffles are actually pea-sized but shown here greatly magnified. Resembling the fallen berries of *Podocarpus* trees, the brightly colored sporocarps are consumed by birds that unwittingly carry and deposit their spores far from where they originated.

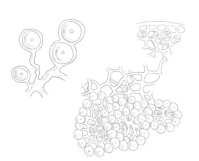

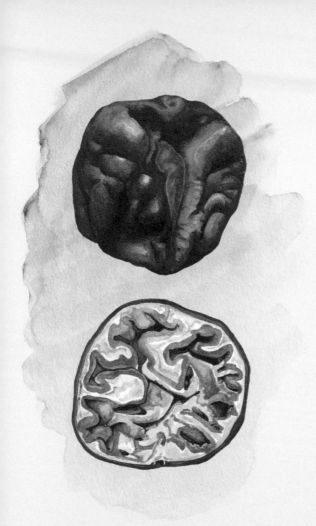

MYCORRHIZAL FUNGI

The vast majority of plant species form a mutually beneficial living relationship with fungi. Mycorrhizal fungi—not roots—are the chief organs of nutrient uptake by land plants. It is likely that the roots of 90 percent or more of the world's plant species (and pretty much all trees) are colonized by symbiotic fungi.

Mycorrhizal (literally, "fungus-root") associations involve fungal hyphae that grow from within and around the roots of the host plant, outward into the surrounding soil, and have the effect of increasing the surface area of the root system several hundred to thousands of times.

OBLIGATE ASSOCIATIONS

Mycorrhizae are so common and fundamental to plant nutrition that, without artificial inputs, most species could not survive without their fungal partners. Essentially benevolent parasites, mycorrhizal fungi benefit from plant photosynthetic products—lipids and carbohydrates—then reward the plant for its hospitality by supplying water and essential nutrients like nitrogen, phosphate, and potassium.

WHEN DID THEY ARISE?

Mycorrhizal associations have been around for about as long as terrestrial plants. The fossil record shows mycorrhizal associations date from 460 million years ago and played a key role in the aquatic plants' invasion of terrestrial habitats. Aquatic plants were unable to survive the harsh conditions on dry land until they joined in symbioses with fungi. Moreover, the earliest land plants, which had no true roots, were colonized by hyphal fungi. These formed structures (for example, vesicles and arbuscules) that were strikingly similar to those produced by modern mycorrhizas. From lowly beginnings, terrestrial plants proliferated—as did mycorrhizal fungi. Indeed, mycorrhizal associations have arisen several times.

↙ Below the surface of any forest, plants rely on symbiotic fungi for water and nutrients; mycorrhizal fungi get the building blocks of life from their photosynthetic partners.

THE WOOD WIDE WEB

An individual plant may have many different species of mycorrhizal fungi connected to its roots at any one time. Likewise, an individual fungus may be connected to multiple plants, including different species. The result is a mycelial network underground dubbed the "Wood Wide Web." Besides transporting water and nutrients, this network functions as a sort of mycelial internet—a communication system where chemical information is shared between plants. Communication signals between plants can stimulate a common defense against soil pathogens, inhibit the growth of neighboring plants, and warn of insect attacks. Nutrition is also shared among plants by way of this common mycelial network. Understory plants and light-deprived seedlings on the forest floor tap into and benefit from this network. For example, the stumps of Pacific Northwest Douglas Fir (*Pseudotsuga menziesii*) felled by loggers continue to live for decades because their roots are fed by those of nearby trees.

ECTOMYCORRHIZAE

Ectomycorrhizal fungi grow into the plant root tissues but do not enter the root cells. Hyphae grow around the outer cortical cells of the root, forming a "Hartig net." Ectomycorrhizas (EcM) exist most often as a mantle or covering of interwoven fungal hyphae on the surface of fine tree roots. The mantle makes the root tips look swollen and is visible to the unaided eye. EcM fungi are associated with most conifers and many hardwoods, including oaks and beeches. Well over 4,000 species of EcM fungi occur in forests worldwide, including many prized edible fungi like boletes, chanterelles, amanitas, and truffles.

ENDOMYCORRHIZAE

Endomycorrhizal fungi are somewhat different physiologically to EcM fungi. This group grows into the plant root tissues and does penetrate the plant root cells. They do not produce a thick mantle over the surface of the root as is common with EcM, nor do they produce showy fruiting bodies. Indeed, most ectomycorrhizae produce no real fruiting body at all; a few species produce balls or clumps of spores in the soil, while many seemingly do not undergo sexual reproduction and may not even have the genes for it.

By far the largest group of endomycorrhizal fungi is the arbuscular mycorrhizal (AM) fungi (phylum Glomeromycota). Arbuscular mycorrhizae take their name from the arbuscules—highly branched structures that they form inside each root cell—where the exchange of water and nutrients occurs. Endomycorrhizal associations involve a much broader array of plants than do EcM. Moreover, there are endomycorrhizal associations unique to specific groups of plants, including alders (*Alnus*), orchids, and ericaceous plants (for example, rhododendrons, azaleas, blueberries, and cranberries). Many of these plants grow in boggy or nutrient-poor soils and benefit from fungal mutualists. In fact, AM fungi are thought better able to scavenge nutrients from the poorest of soils, including rocky and arid ones. In addition to conferring drought tolerance, survivability in nutrient-poor soils, and other benefits, the Glomeromycota are crucial for building and maintaining soils.

Most plants, including grasses and cereals, vegetables, vines, and bushes, and even those that do not form associations with EcM fungi are known to partner with AM fungi. There are quite a few plants known to form mycorrhizal associations with both AM and EcM fungi, and more are being discovered all the time. Maybe most plants go both ways!

CRYPTIC, ENIGMATIC FUNGI

Most endomycorrhizal fungi are poorly known, given their cryptic nature, and most cannot be cultured in the lab. Ironically, what is known about them is that they dominate the planet and are probably the puppet masters for all life on terrestrial earth. How do we know this? Although we cannot see many or even most of the species out there, they leave their DNA behind in the soil, which scientists can see.

↓ This root cross-section shows arbuscular mycorrhizal fungi within plant root cells. From there, the fungi grow out into the soil and bring water and nutrition to the plant host.

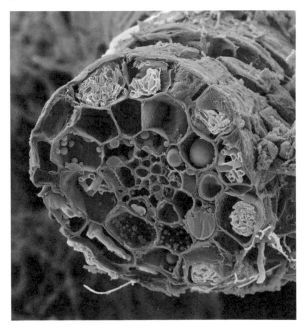

LICHENS

L ichens are mutualistic fungi. In contrast to mycorrhizal fungi, which maintain their hyphal appearance, lichens do not immediately appear to be fungal. They more closely resemble small plants, and were, in fact, mistaken for plants until the second half of the 19th century. We now know that lichens are comprised of a fungus (mycobiont) and a photobiont, either an alga or cyanobacterium, or both. Much of the entire structure is fungal and functions to acquire nutrients and to house the photobiont. The role of the photobiont is crucial in that it produces carbohydrates via photosynthesis.

Many lichens thrive in extreme environmental conditions where no other photobionts can. They can grow on almost any substrate you can imagine: tree bark, rocks, and soil, of course, as well as buildings and gravestones. Some ecosystems such as the tundra, Antarctica, and coastal fog deserts are dominated by specific lichen communities and few other organisms are able to exist under the conditions there.

↓ A lichen is a community of photosynthetic organisms, usually algal cells (A and C) protected within a structure made of fungal cells (B, D, and E).

The photobionts photosynthesize when conditions are favorable, and support all the symbionts with carbohydrates.

The fungal tissues resist desiccation and hold fast to surfaces by short, root-like structures, or rhizines (F), underneath.

→ In some habitats, it is common to see surfaces covered with many different otherworldly organisms known as lichens. Each lichen is composed of a fungus (the mycobiont) and a photobiont, and the reproductive structures often resemble those of the mycobiont involved—in this case, an ascomycete cup fungus.

MYCOTOXINS

Most molds are not harmful, but some produce powerful toxins called mycotoxins. Mycotoxins are those specifically produced by fungi (like *Fusarium*, *Alternaria*, *Aspergillus*, and *Penicillium*). Not entirely understood, some seem to kill host cells or disrupt the host plant's immune response, while other toxins likely kill competing microbes in their habitat. Alternatively, mycotoxins may be used for communication between like species and just so happen to be toxic to other life.

FUSARIUM

Pathogens belonging to the genus *Fusarium* are the causal agents of the most significant crop diseases worldwide. Virtually all *Fusarium* species synthesize toxic secondary metabolites, known as mycotoxins. The management of the *Fusarium* phytopathogens has proven to be difficult due to their high genetic variability and broad host specificity. The species can adapt to a wide range of habitats, including the tropical and temperate areas, and are considered as one among the most devastating plant pathogens. *Fusarium* colonizes the living plant host, then produces toxins and cellulolytic enzymes, hijacking the host's metabolic pathways and nutrients. Mycotoxin-producing *Fusarium* species are major pathogens in cereals like wheat, oats, barley, and corn. Also, they can cause up to 50 percent yield loss in tropical fruit crops like banana and pineapple, as well as lentils, tomatoes, and peas.

→ *Fusarium* species are common plant pathogens and food contaminants. Seen under the microscope, their canoe-shaped conidia (asexual spores) afford easy identification.

Fusarium species produce some of the most dangerous naturally produced substances on the planet. These substances are dangerous, carcinogenic, and even deadly to humans and livestock. Animals consuming plants contaminated with *Fusarium* may end up as collateral damage. (They are so toxic to humans that some have been produced intentionally as bioweapons.) Many toxins like fumonisins and trichothecenes are heat-stable and cannot be deactivated by cooking. The only way to surpass this situation is by preventing or inhibiting the production of mycotoxins in the field.

ASPERGILLUS

Aflatoxin, produced by the fungus *Aspergillus flavus*, is the most carcinogenic substance naturally produced on Earth. All corn, peanuts, and some other grain are screened to ensure this dangerous mold is not present. While the toxin is potent, even in trace amounts, it is easily detected since it fluoresces under ultraviolet light.

ENDOPHYTIC FUNGI

I n just about every plant group investigated, there seem to be species of endophytic fungi living in association with plant hosts. Likewise, endophytic fungi appear to play key symbiotic roles in the lives of their plant hosts, providing drought tolerance through plant-like hormones or by affording protection from mammalian and arthropod herbivory due to the production of toxic compounds.

Scientists studying the relationship between endophytes and their hosts are discovering novel chemical compounds of great importance. As one example, the cancer "wonder drug" paclitaxel (brand name Taxol), which was discovered in rare and slow-growing Pacific Yew (*Taxus brevifolia*), seemed likely to doom the trees because they had to be killed in order to obtain the drug. It was discovered that the source of the compound was actually an endophytic fungus living within the trees. Better yet, the fungi can be grown in culture (plus the drug can be manufactured semi-synthetically) and the trees have been spared.

↓ Many fungi live as endophytes (meaning "within-plants"), their hyphae growing between the cells within plant tissues and seen here with a microscope.

↓ Plant tissue examined under a microscope in cross-section will reveal endophytic fungal hyphae growing around the plant cells.

→ The study of the interactions between endophytic and epiphytic fungi, and their hosts, may lead to the discovery of novel chemical compounds. For example, the cancer "wonder drug" paclitaxel (PTX) was discovered in the Pacific Yew, which is found in Pacific Northwest forests and also grown horticulturally.

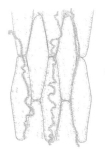

MYCORRHIZAL THIEVES

Mycorrhizal fungi are benevolent parasites. They have a parasitic lifestyle, but at the same time benefit their host plant. Undoubtedly, mycorrhizal fungi evolved from parasitic ancestors. That a symbiont can shift from a parasitic to a mutualistic relationship with its host over evolutionary time is expected; sometimes a symbiont may even be mutualistic or parasitic at different phases of its life cycle or that of the host. In most cases, the host is the photobiont. But not in all cases. Turning this line of thinking on its ear, some mycorrhizal plants turn the tables and are parasites of mycorrhizal fungal symbionts.

GHOST PIPES

Monotropoid plants show that parasitism can go both ways. Ghost Pipes (*Monotropa uniflora*) have no chlorophyll and cannot photosynthesize. They are "mycoheterotrophs"—that is, plants that parasitize mycorrhizal fungi. Thus, they steal photosynthate from surrounding plants by tapping into those plants' fungal partners in the soil. This is known as "epiparasitism." It is assumed that mycoheterotrophic epiparasites like *Monotropa* must be giving something back to their fungal partners, but this is still not certain; it seems totally unlikely that they are not giving anything at all to the photosynthetic plant symbionts.

← *Monotropa uniflora*, known as Ghost Pipes, are achlorophyllous (meaning they cannot photosynthesize). This is because their flowers and greatly reduced leaves are no longer useful for catching light.

There are non-green plants. Chlorophyll makes plants green and, more importantly, is a sign that they are photosynthetic autotrophs. But what about plants without chlorophyll: where do they get their energy? It was long assumed that achlorophyllous plants like Ghost Pipes cannot photosynthesize and therefore must do something else. Some, like dodder (*Cuscuta* species), are parasitic on other plants. Or might they be saprophytes—obtaining their nutrition from decaying organic matter? It wasn't until the 1960s, when radioisotope experiments demonstrated the movement of carbon from spruce trees to *Monotropa*, that it became clear fungi were involved in this carbon flow. Scientists were astounded to see that these plants were stealing carbohydrates from other plants and had unwitting accomplices in their crimes: mycorrhizal fungi.

ORCHIDS

Orchid mycorrhizae also show that parasitism can go both ways. Orchids function in much the same way, getting their sustenance from mycorrhizal fungi. While many orchids do photosynthesize, there are many achlorophyllous species that do not. The association between orchid and fungus is very highly evolved; orchids do not make true seeds with a nutrient source for the baby plant (endosperm) like other flowering plants. Instead, orchid seeds are merely naked embryos and very minute, about the size of a grain of dust. Orchid "seeds" must be parasitized by their specific mycorrhizal fungus in order to begin germination; the fungal mycelium is the only "root" the young plant has and the source of all its nutrition.

WHY DON'T THE THIEVES GET CAUGHT?

Plants are adapted to allow infection by a large number of mycorrhizal fungi and seem perfectly willing to allow the net flow of carbon to other plants, as pointed out above. So it seems they are ill-equipped to detect cheaters in this system. As long as the epiparasitic plant does not end up compromising the fitness of the fungus, the long-term stability of their food source is assured. The ancestors of mycoheterotrophs, like most plants, are likely to have been mycorrhizal; the loss of chlorophyll and the resulting cheating of the common mycorrhizal network came later on.

THE HISTORY OF MYCOLOGY

Beginning with Aristotle, all living things were treated as either plants or animals, simply according to whether they could move or not. Today, with a much better grasp of the evolutionary relatedness of all life on the planet, it turns out that the organisms most closely related to animals (indeed us!) are . . . fungi.

Italian Pietro Antonio Micheli (1679–1737), considered the founder of mycology, started out as an apprentice bookbinder and read works by the great botanists of the day. Swedish Carl Linnaeus (1707–1778) is best known for inventing the binomial nomenclature system, which consists of a generic name followed by a specific epithet, and taxonomy based on sexual features.

DOWNY MILDEW SOLVED BY
SCIENCE . . . AND SERENDIPITY

Some of the greatest scientific discoveries come about by serendipity—being in the right place at the right time—and astute observation. In the late 19th century, vineyards in France were plagued by a disease called Downy Mildew of Grape, caused by the oomycete *Plasmopara viticola*.

In 1876 the French botanist Pierre-Marie-Alexis Millardet happened to be strolling home past local vineyards when he noticed that the grapevines nearest the road were splashed with a strange, blue-green-colored substance. His curiosity piqued, he inspected the plants, noting that the leaves were completely free of Downy Mildew—but only wherever this blue-green substance had been applied. The grower revealed that a mixture of copper sulfate and lime had been applied to the plants to discourage pilferers from picking his grapes. Millardet found this "Bordeaux Mixture" worked well against all manner of fungi. A century and a half later, the Bordeaux Mixture remains the most-used fungicide to this day.

By the early 19th century, fungi were placed in a Fifth Kingdom. Swedish Elias Magnus Fries (1794–1878) unified mycologists of the world with a system based solely on macroscopic morphological characters (white-spored mushrooms grouped together, dark-spored mushrooms, boletes, polypores, and so on), which continues to be an indispensable aid for anyone preparing for the study of higher fungi.

MYSTERY OF DUTCH ELM DISEASE SOLVED

American Elm (*Ulmus americana*) has long been the preferred tree of urban planners and urban foresters; its perfect form, heavy shading foliage, very high spreading canopy, and lack of messiness made it the perfect street tree of cities in eastern North America and elsewhere. And so it was that American streets were lined with them; city parks and college campuses were forested with them. But that all changed in the early 1900s when a strange disease began killing them.

In 1921 the mystery of elm tree death was solved in the laboratory run by the Dutch plant pathologist Johanna Westerdijk. The fungus was identified by careful observation of infected trees and careful cultivation in the lab by one of Westerdijk's graduate students, Marie Beatrice "Bea" Schwarz. Growing the mold out of infected wood and inoculating a healthy tree with it quickly caused symptoms of the disease in that tree, and death soon followed. The re-isolated fungus was an asexual mold, which was dubbed *Graphium ulmi* in 1922. The sexual stage was later discovered and named *Ceratostomella ulmi* by Christine Buisman, also of Westerdijk's lab (this fungus was subsequently named *Ceratocystis* [*Ophiostoma*] *ulmi*).

← Removing the bark from a dead tree reveals severe damage to the vascular tissue caused by bark beetles—the insects are the vector of Dutch Elm Disease fungal pathogens.

HIDDEN LIFE

Fungi may remain underground or within a plant host for years or even centuries, but they are there, doing what fungi need to do to survive. The mycelium is the main body of the fungus and it handles all the physiological aspects. Unless the fungus is in reproductive mode, and making showy fruiting bodies, the mycelium mostly remains hidden from us. Rest assured, the fungi are there, although we often don't see them.

HIDDEN "CRYPTIC" SPECIES

Taxonomists have long suspected that many undiscovered species of fungi are hidden. Are they disguised in some way? Sort of . . . many lie hidden within known groups of very similar looking species. Called species complexes or cryptic species, these fungi are morphologically indistinguishable from—and share a single scientific name with—other species but are, in fact, phylogenetically distinct. Species complexes are commonly seen among plant pathogenic fungi. Many macrofungi—including groups familiar to us—have long gone by the names of European species that look identical, or nearly so. This includes all sorts of boletes, amanitas, russulas, and pretty much most other groups of mushroom-forming species.

Modern DNA sequence analysis has unmasked hidden species in many basidiomycetes and ascomycetes, which make up the bulk of all fungi. It is predicted that the number of novel species hidden within complexes is likely around 1.4 million and, once discovered, they alone will increase lichen totals eleven-fold and the known number of fungi five-fold.

MUSEUMS, HERBARIA, AND FUNGARIA

For several hundred years, plants and fungi have been collected and deposited in museums, herbaria, and fungaria. These collections are invaluable to taxonomic research, since they provide a time capsule of life from a given region at a given period of time. Many of these collections probably house untold species unknown to science. We know this because many of the newly described species each year come not from some unexplored corner of the world, but some understudied drawer in a museum cabinet. Unfortunately, collections continue to grow but the number of mycologists (and taxonomists in all fields) is not growing quickly enough to keep apace.

↓ Museums, herbaria, and fungaria are invaluable repositories for scientific specimens of organisms. London's Kew Gardens, shown here, has the largest fungarium in the world.

THE AGE OF CITIZEN SCIENCE

It seems that everywhere you turn in science and natural history circles, the term "citizen science" is being used. The simplest definition of citizen science is the one used by the National Geographic Society: "Citizen science is the practice of public participation and collaboration in scientific research to increase scientific knowledge. Through citizen science, people share and contribute to data monitoring and collection programs."

Citizen scientists have long done much of the work needed for observations, documentation, and collecting data in many scientific fields like astronomy, meteorology, botany, and zoology—especially so with birds. Citizen science projects today are singularly popular—and crucially important—in the fields of ecology and conservation biology. Currently, mycology is especially in need of help from dedicated amateurs. There are insufficient professional mycologists, so utilizing members of the public (for example, in mushroom clubs and societies) will be ever-important.

TRADITIONAL VERSUS MODERN OBSERVATIONS

The traditional practice of amateur mycology involved collection of specimens, identification, creation of species lists, and drying/preserving specimens. This is similar to the model of amateur birders who create species lists at annual Christmas bird counts. For citizen mycology in the 21st century it has been deemed absolutely necessary to add two components: depositing preserved specimens in established fungaria and obtaining genetic information by sequencing DNA.

← Morphological characters of the fungus are useful for identifying it, but some characters are very small and require magnification using a hand lens or microscope.

WHAT IS DNA BARCODING?

DNA barcoding utilizes a single, short section of DNA sequence to identify species—plant, animal, fungal, and so on. Unique to barcoding is the idea of using short DNA sequences from a single, standardized region of the genome to identify species from a wide taxonomic range across kingdoms. DNA is extracted from organisms (alive or long since dead), pathogens living in host tissue, or even extracted from soil samples. The DNA is analyzed and sequences of the barcoding target region are obtained from the specimens. For most organisms, the DNA barcoding region is a portion of one gene, comprising around 650 base pairs (the As, Cs, Gs, and Ts of the stringy DNA molecule) from the first half of the mitochondrial CO1 gene. The resulting DNA sequence data are then used to construct a phylogenetic tree with related individuals clustered closely together. DNA barcode sequences vary extensively between species but hardly at all within them, thus can be used to distinguish one species from another. As a result, when we look at the resulting phylogenetic tree, each cluster of closely related individuals that we see is assumed to represent a separate species.

Documentation, including photographs, of morphological and ecological data of collections has always been essential. Online repositories for observations include iNaturalist (probably the most popular of late), Mushroom Observer, Flickr, Myco Portal, and Discover Life. Vouchering of collections has also been important historically. For long-term preservation, curation, and availability to scientists, specimens should be kept in professional herbaria and fungaria. New to the 21st century, a DNA sequence for the vouchered organism is also often kept. Traditionally, DNA sequence analysis was performed solely at a university where sophisticated (and expensive) equipment could be managed. Nowadays, non-academic scientists are also getting in on the act. And new techniques for DNA sequence analysis, such as DNA "barcoding," make the entire process cheaper and more efficient.

CULTIVATION

The hot new trend? Growing edible mushrooms at home! People have historically been growing mushrooms right along with their fruits, vegetables, and livestock for centuries. Both rewarding and sustainable, you can use lawn wastes, kitchen scraps, and even newspaper and cardboard waste.

Some types of mushrooms are biotrophic (needing a host to stay alive) and cannot be cultivated. Many others are saprobic in nature—they decompose organic material—and those can be cultivated. Many wild mushrooms found in the fields and woodlands across most of the globe, like field and oyster mushrooms, have already been domesticated. Other species, like Shiitakes and Namekos, once curious exotic mushrooms in restaurants, are now commonplace on grocers' shelves. Even if you don't enjoy eating them, you can still get a lot of enjoyment from cultivating mushrooms—they're fun to observe, beautiful to photograph, and those fungi are always at work for you, creating rich soil from wastes that you might otherwise send off to landfill.

↓ Mushroom-growing kits have become popular. Lion's Mane mushrooms (*Hericium* species) are delicious, healthy, and easy to grow right out of a box.

↓ Shiitake mushrooms have long been grown on oak logs but nowadays it's become commonplace to grow them indoors on inoculated sawdust.

→ Mushrooms that are mycorrhizal partners of trees and other plants cannot be cultivated, but many saprobic wild mushrooms have been successfully domesticated. This includes oyster mushrooms (*Pleurotus* species), a popular culinary mushroom.

EXTINCTION

White Ash trees (*Fraxinus americana*) are widespread across much of eastern North America and are the host of the Ash Bolete mushroom, *Boletinellus* (syn. *Gyrodon*) *merulioides*. This mushroom is commonly seen in the late summer and fall on lawns and in parks, as well as in woodland habitats. The mushrooms often fruit in large numbers and always in close proximity to their host tree but are not normally of much interest to anyone.

Sadly, ash trees are in decline in parts of North America, falling victim to an invasive beetle—the Emerald Ash Borer (*Agrilus planipennis*). And as the trees decline—potentially going extinct—so goes this marvelous mushroom. The Emerald Ash Borer was first seen in the Detroit, Michigan, area in 2002. The adults, iridescent green beetles about the size of a grain of rice, feed on the tree's leaves and lay eggs on the bark. The larvae kill ash trees by burrowing through the bark into the phloem tissues that transport water and nutrients. The borer has attacked and killed tens of millions of trees in at least 35 states, mostly in the eastern and central United States; it has also infested southern Canada. By 2017 the International Union for Conservation of Nature (IUCN) declared the borer had caused six North American ash species to become endangered or critically endangered.

RELATIONSHIP STATUS: COMPLICATED

Long considered a mycorrhizal bolete, the phylogeny of *Boletinellus* was uncertain, with this group of mushrooms shuffled among various taxonomic groups. About the only thing certain about this fungus was that it was mycorrhizal, just like all boletes were thought to be. It turns out that none of this was true. Upon close inspection, it seems the mushroom is actually a symbiont of an aphid that lives as a parasite on the roots of the tree. The fungus seems to afford the tiny bug some protection by growing around the insect and forming dark black galls on the roots of the host trees. That's right: the aphid is inside the hyphal galls, while feeding on the tree—the fungus seemingly getting all its nutrition from the insect.

← The Ash Bolete is unusual in every way. From above it looks like any other bolete, but the underside features a bizarre merulioid, or veiny hymenium.

DETECTING INVASIVE SPECIES

The common Fly Agaric (*Amanita muscaria* var. *muscaria*) from Europe and Asia is spreading rapidly around the world, even turning up in North America. (Some red North American Fly Agarics look similar but are not the same.) Scientists studying invasive fungi like the Fly Agaric of Europe have determined that it is especially widespread in South America. Why is this?

To meet a global demand for timber, tree plantations were planted in South America during the early 20th century. Extensive plantings of non-native species are now found in Brazil, Chile, Argentina, and Uruguay. When you plant tree seedlings, you get the tree's mycorrhizal fungal partner too—hitchhikers on the roots. In Colombia, plantations were created in the 1950s, during a period of intensive logging when policies to limit deforestation in native forests were established. One unforeseen consequence of planting non-native trees was the introduction of non-native fungi. In Colombia the Fly Agaric has become established and spread to local oak forests.

↓ Conifer trees are intensively used for lumber and pulp, globally. Where the seedlings go, their fungal symbionts travel with them.

↓ *Amanita muscaria* is a very common symbiont of many tree species, including conifers, and it is key to their successful growth.

→ Possibly the most recognizable mushroom on the planet is the Fly Agaric (*Amanita muscaria*). Although not deadly, it is a commonly encountered and toxic mushroom, with varieties native to North America, Europe, and Asia. It is now found on all continents except Antarctica, having been moved around due to human activity.

FUNGI AS FOOD

Fungi are used to create all sorts of fermented foods, beverages, and flavorings. As long as there are simple sugars present, fungi can ferment them to alcohol—for example, fruit juices are fermented to make wine. Starchy grains are not fermentable and must be sprouted first; the seedling creates amylase enzyme, which converts starch to sugar for use by the baby plant. Dried and roasted, this malt is now ready for brewing into beer (whiskey is essentially made from distillation of beer).

INDUSTRIAL WORKHORSES: *ASPERGILLUS* AND *SACCHAROMYCES*

Since saké is made from grain, in this case rice, it is a beer. However, making saké requires two fungi to be added to cooked rice. *Aspergillus oryzae*, also known as "kōji mold," produces copious amounts of amylase enzyme that breaks down the rice starch to a fermentable sugar. *Saccharomyces cerevisiae*, a brewer's yeast, carries out the fermentation. *Aspergillus oryzae* is a workhorse of Asian cuisine, used to make miso, soy sauce, and vinegars, as well as countless other fermented bean pastes and sauces.

↙ *Aspergillus* species conidia (asexual spores) and conidiophores. This structure is reminiscent of an aspergillum, the holy water sprinkler tool used by Christian priests.

Another *Aspergillus*, *A. niger*, is used in industry to make many useful enzymes and other products, including high-fructose corn syrup. The most economically important product of *A. niger* is citric acid, a popular flavoring in many foods and soft drinks. It is far cheaper and easier just to grow any number of fungi that excrete citric acid as part of their metabolism.

NOBLE ROT

Botrytis cinerea is a ubiquitous food spoilage mold and probably ruins more refrigerated fruits and vegetables worldwide than any other microbe. Given enough time, this fungus can and will spoil any piece of fresh fruit in your home. It's a serious pest to growers of many crops. Grapes can be attacked by this fungus, if still on the vine late in the season. But under the right conditions, the grapes are magically transformed by "Noble Rot." How does the fungus work its magic? The fungus pierces the grape skins during infection. This allows moisture to escape and the infected grapes shrivel into raisins. The water loss concentrates sugars and flavors; the flavors are further transformed by the Noble Rot fungus. Sauternes wines of France's Bordeaux region and Tokays of Hungary have been made this way for centuries. Thus, two molds are needed to make Sauternes wines: regular *Saccharomyces*, plus *Botrytis*.

A UNIQUE CHEESE

The remoteness of Iceland has resulted in foods that are unknown elsewhere. Sheep fat, rather than milk, is used to make a sort of blue cheese. Called *hnðmör* (kneaded fat), it is made in much the same way as dairy cheese elsewhere in the world. Sheep fat is pressed and placed in a mesh cloth, then hung in a cool place, usually some sort of shed. After four or five days, a greenish mold will appear on the fat. The mold and fat are kneaded together and then cut into blocks that look very much like Roquefort cheese before being left to age in a refrigerator or cellar. The mold seems to be a sort of *Penicillium* species, probably closer to *P. roquefortii* or *P. camembertii*. Not surprisingly, cheeses like Gorgonzola and Roquefort, as well as soft cheeses like Camembert and Brie, are ripened by *Penicillium*.

SUSTAINABLE MYCO MATERIALS

Down through the ages, humans have foraged for all manner of organisms in nature for food, fuel, and fiber. Lately, fungi are showing promise due to the increasing need for environmentally sustainable materials. Fungal-produced materials include mycelial biocomposites, packing foams, and textiles.

"Myco-board," a wood-like material made from organic debris left over from agriculture and some other industrial processes, is made of densely packed mycelium. After the substrate is cooked, moistened, and inoculated, the mixture is poured into a mold of the desired shape and left to grow in the dark. A week or two later, the finished product is popped out and the material rendered biologically inert. Ecocradle®

is similarly produced and is an alternative to Styrofoam™ as a packing material for shipping fragile goods. Unlike Styrofoam™—which is hard to recycle and not biodegradable—this myco-material can easily serve as mulch in your garden. Using the same technology, and fungi, sustainable "myco-insulation" is coming to the home-insulation market and can be made fire-retardant.

WEARABLE FUNGI

On the way out: fur, feathers, animal hides, and petroleum-based synthetic leather. Now trending: sustainable fungal-derived "myco-leather." This is made from sustainable substrates (like agricultural waste, paper pulp waste, or sawdust) and when it's worn out, it can be composted. Mycelium-based leather is produced by polypore fungi; these wood rotters are well known for their ability to degrade lignocellulosic material (some have long been used in the pulp and paper industry or in chemical production). The cells walls of polypore fungi are composed of a matrix of chitin, glucans (glucose polymers), and glycoproteins, and thus form a flexible cellular architecture similar to animal hide leather.

To manufacture the leather, wood rot fungi are introduced to a woody substrate. Decomposition ensues and the resulting material is then processed using chemical and heat treatments, making the leather-like fabric extremely durable and resistant to environmental stress; plasticizing and crosslinking of mycelial mats with various reagents creates a leather-like final product. Since you were going to ask: no, myco-leather cannot begin growing if it gets wet in your closet or on your person—the mycelium-based material is heat-killed after growth!

← Exciting new fungal fashions may soon be coming to a runway or department store near you! *Ganoderma* fungi are often the starting point, with a final product very similar to leather.

FUNGI AS CAUSE AND
CURE OF DISEASE

Penicillin, a compound excreted by *Penicillium* mold, was the first antibiotic and has saved untold lives. The discovery of penicillin was purely serendipitous—in 1928 Alexander Fleming noticed a contaminant mold growing among a bacterial culture. Furthermore, there appeared to be a clear "halo" surrounding the mold. Bacteria could grow to that zone but something in the culture medium prevented their growth from getting any nearer to that mold. Fleming reasoned that the fungus was excreting something into the agar medium. He was correct and isolated the substance, naming it penicillin.

It wasn't until 1940 that two other researchers, Howard Florey and Ernst Chain, "rediscovered" Fleming's experimental notes and were able to create a stable form of penicillin that could be administered orally to a sick patient. Although many other researchers were, of course, involved in what has become one of the greatest discoveries for humanity, Fleming, Chain, and Florey shared a Nobel prize for their discovery.

Antibiotics work in seemingly miraculous ways because they target physiological pathways in bacteria that are not possessd by animals. Thus they don't typically have any effect on human cells. (Their amazing utility has led to overuse and recently some bacteria have evolved resistance to these medicines, rendering them useless against certain pathogens.)

MIRACLE DRUGS

Many fungal-derived antibiotics have been discovered, including cephalosporin and griseofulvin; numerous semisynthetic penicillins are also common pharmaceuticals (methicillin, ampicillin, carbenicillin, amoxicillin, and so on). The miracle drug cyclosporine is an immuno-suppressor given to a patient receiving an organ transplant to avoid transplant rejection. This drug was originally derived from *Tolypocladium inflatum* (the anamorph of a *Cordyceps* relative, which is parasitic of truffles and insects).

BLOCKBUSTER DRUGS

The statin Lipitor® (atorvastatin) is one of the most commonly prescribed drugs in the world. Statins reduce the risk of heart attacks and strokes. They work by lowering the level of cholesterol in the blood and have anti-inflammatory effects, thus preventing blood clotting and buildup on blood vessel walls. Common statins today include atorvastatin, fluvastatin, simvastatin, and rosuvastatin. Nowadays produced synthetically, statins were originally isolated from species of *Aspergillus* and *Penicillium*.

MYCOSES ARE DIFFICULT TO TREAT

Fungi are very different from bacteria; evolutionarily they are much more closely related to us. Since our physiology is quite similar to that of fungi, fungal pathogens can be quite difficult to treat. The trick is creating poisons that kill fungi but not our own cells. Even a simple toenail fungus can take a long time to vanquish and the number of therapies on the market are limited.

Human mycoses are on the sharp uptick and more than 1 million people die of fungal infections every year. The usual suspects are species from just four genera: *Aspergillus*, *Candida*, *Cryptococcus*, and *Pneumocystis*. Although great strides were made in the 1990s, drug development has largely stalled since then. Opportunities exist for accelerating development, particularly in fungal asthma, and to treat chronic and invasive aspergillosis.

→ *Penicillium* species growing in culture. These ubiquitous fungi can grow on a wide range of substrates. All too frequently, this beneficial fungus is a contaminant in labs.

FUNGI AS MEDICINE

The genus *Psilocybe* is a large group of small, brown mushrooms that grow on decaying wood or the dung of mammals. These mushrooms contain the psychotropic tryptamine compound psilocybin, or its analog, psilocin. Both resemble the neurotransmitter serotonin, structurally, and as a result bind with and activate serotonin receptors in the brain.

Psychedelics produce altered states of consciousness characterized by changes in perception, cognition, and mood. It has long been recognized that these compounds may have therapeutic potential for neuropsychiatric disorders, including depression, obsessive-compulsive disorder, and addiction. Indeed, psilocybin and psilocin were used to successfully treat tens of thousands of patients in the 1950s and '60s. Both have recently returned to the forefront of research and show much promise as therapeutic drugs. Among psychedelics, psilocybin has recently been shown to relieve depression symptoms rapidly and with sustained benefits for several months— all after a single dose of the drug.

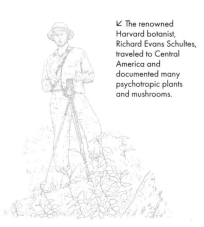

↙ The renowned Harvard botanist, Richard Evans Schultes, traveled to Central America and documented many psychotropic plants and mushrooms.

→ Prior to 1957, very few people had heard of the small mushrooms belonging to the genus *Psilocybe*. That all changed on May 13, 1957, with the publication of an article in *Life* magazine by Gordon Wasson. His amazing story (along with grainy photos) told of mystical ceremonies and the ritual use of hallucinogenic mushrooms by people in southern Mexico.

FUNGI AND POPULAR CULTURE

Classical music, poetry, and literature have often featured mushrooms. Nineteenth-century Russian composer Modest Mussorgsky's "Gathering Mushrooms" comes to mind. Pulitzer Prize-winning poet Paul Muldoon features mushrooms in several poems, including an ode to "Destroying Angel."

Several novels with tie-ins to mushrooms (usually amanitas) have been published over the years. They mostly involve a murder mystery and often have a predictable (and mycologically-inaccurate) storyline. A good whodunnit is Sue Grafton's *I is for Innocent*. The murder weapon? A strudel pastry with Death Caps baked right in.

Possibly the best work of myco-fiction was *The Purple Pileus* written by sci-fi master H. G. Wells. Wells describes how the story's milquetoast protagonist eats some wild mushrooms that he finds one day and has a transforming experience. The physiological effects of the mushrooms are practically identical with those produced by an ingestion of Fly Agaric (*Amanita muscaria*). The protagonist goes berserk and rages at everyone around him—stopping just short of killing anyone, he ends up with life-changing courage. All that from a single dose of ibotenic acid!

Most animalian victims that fall prey to fungi are pretty small—for example, insects and other arthropods, and nematodes. What happens when *Cordyceps*-like fungi finally evolve the ability to attack larger quarry, like us? That's the premise of the popular HBO series *The Last of Us*, where a fungal outbreak ends up killing off almost every human being in sight. It probably can't happen—for now.

REAL-LIFE ZOMBIE FUNGI

Ants parasitized by the fungus *Pandora formicae* move away from the colony. In most instances, the host is compelled to crawl up and latch onto the substrate (an act known as "summit disease"). A macabre death follows and the fungal sporophores then erupt through the host's exoskeleton and spores are launched. Ants parasitized with *Ophiocordyceps unilateralis* position themselves on foliage directly above paths frequented by other members of their colony—infectious spores then rain down upon the unwitting next-victims below. Many of these fungi have been known for a long time, but it wasn't until they were dubbed "zombie fungi" that they grabbed headlines.

← There are many fungal pathogens of animals. One very bizarre group of pathogenic fungi, the "zombie fungi" (*Cordyceps* species), specializes on insects and other arthropods.

MYCOTOURISM

Mycotourism is a type of tourism centered on fungi exploration, including mushroom foraging and educational experiences. It involves guided walks for mushroom collection, culinary events featuring locally foraged mushrooms, and educational programs on fungal ecology and medicinal properties.

Mycotourism also highlights the cultural significance of mushrooms and promotes conservation awareness. Fungal tours, workshops, and events provide insights into mycology and sustainable foraging practices, fostering a deeper connection between participants and the natural environment.

The largest North American wild mushroom festival is the Telluride Mushroom Festival, which originated from a conference started by Dr. Emanuel "Manny" Salzman in Aspen, Colorado, in 1977. Initially focused on mushroom poisoning and psychedelics research for physicians, it has grown to encompass all mycological topics. For decades the face of the festival was the late NYC mushroom expert Gary Lincoff.

↓ Today, the Telluride Mushroom Festival features mushroom-themed events, wild mushroom dishes in restaurants, lectures by world-renowned mycologists, expert-led forays, and the popular mushroom parade.

→ More and more people are foraging for wild mushrooms for culinary use. At the end of the foray, lucky mycophiles can be treated to a wide array of colors, textures, and flavors. But it is imperative to know what you are doing—some mushrooms are deadly poisonous.

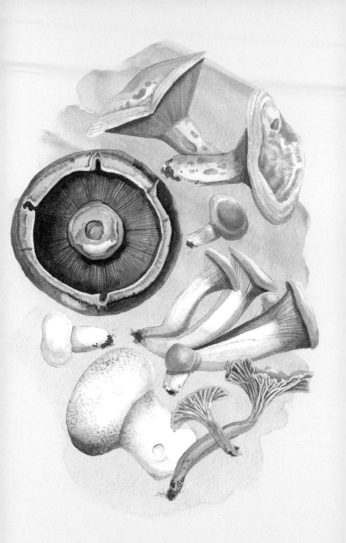

A CHANGING CLIMATE

Global climate change and its effects have been studied for decades. The geographic ranges and habitats of species are changing. Some places are becoming too hot or dry (or wet). Other places that were once too wet, too dry, or too cold are now becoming more favorable for some. Of course, climate change will leave no favorable habitat at all for some species that will inevitably go extinct.

FUNGI AND CLIMATE CHANGE

A changing climate has led to other observations. Some plants are flowering earlier and some are now blooming twice in a season. Fungi are mostly hidden throughout the year and harder to study than plants, but seem to be following the same patterns. Mushroom fruiting times are happening earlier in the year with some species— still others are fruiting twice per year. The rise of social media has allowed us to see such observations globally and in real time.

CARBON IS THE PROBLEM

The global climate has been warming for a long time but humans have drastically accelerated this through the burning of fossil fuels—this pumps tons of carbon waste into the atmosphere. There is a movement afoot to fight global climate change head on, and one of the most powerful tools just might be fungi.

~ Fungi are the solution ~

Arbuscular mycorrhizal fungi (AMF) are poorly known—just about the only thing we know about them is that they are ubiquitous all over the planet and seem to partner with most plant life. Moreover, they form a symbiotic relationship with the majority of our important crop species. Recently, scientists are coming to the conclusion that an effective way to pull carbon dioxide out of the environment, and at the same time increase our crop plant production, is to employ agricultural practices that favor these beneficial soil fungi.

AM FUNGI: THE KEY TO
SOLVING CLIMATE CHANGE

It is now well understood that intensive agricultural practices reduce the good microbial populations in soils and the quality of soils overall. Other conventional agricultural practices reduce or eliminate mycorrhizal activity in the soil and release carbon dioxide into the atmosphere. AM fungi might just be the key to the problem of global climate change. Left to do their thing, AMF produce glomalin, which can last for decades or centuries in undisturbed soil.

GLOMALIN: THE MOST IMPORTANT MOLECULE

AM fungi dramatically increase the effective root systems of plants by producing a vast network of nutrient- and water-absorbing hyphae, which greatly increase the surface area of a plant's roots. But these fungi are important in one other way: AM fungi and associated soil microorganisms produce a sticky protein called glomalin, which can catalyze carbon sequestration and storage in soil. Glomalin acts as an organic glue that creates a soil architecture that permits air, water, and roots to easily move through it, so preventing soil erosion. In addition, glomalin creates stable soil aggregates that protect soil carbon. Glued together by glomalin, soil aggregates shelter organic matter rich in carbon and nutrients. Furthermore, glomalin may account for as much as one-third of the world's soil carbon and the soil contains more carbon than all plants and the atmosphere combined.

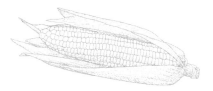

← Arbuscular mycorrhizal fungi are crucial to the planet. They form a symbiotic relationship with the majority of our important crop species, like corn.

INVASIVE SPECIES

Many invasive species are adaptable and seem to be benefiting from a warming environment. Recent research has shown that some invasive species are able to complete their life cycles and reproduce more quickly with warmer temperatures.

Most people don't regard mushrooms as potentially invasive. Yet the infamous Death Cap (*Amanita phalloides*) seems to be spreading across the globe. Likewise, the European Fly Agaric (*A. muscaria*) appears to be moving with some timber species grown on tree farms and plantations. In eastern North America we are seeing the Golden Oyster mushroom (*Pleurotus citrinopileatus*) naturalizing and spreading in some forests.

The concern is that these invasive species may naturalize and outcompete other native fungi, with unknown effects on native forests. There have been few success stories—once established, invasive species are very difficult to remove. We cannot turn back the clock on many, or even most, invasives, but educating and involving the public can help keep many of these pests in check.

↑ The Golden Oyster mushroom is tasty and easy to cultivate, which is probably how it arrived in eastern North America from Asia, but it is aggressively saprobic, rotting just about any dead wood available.

↑ The Orange Pore fungus (*Favolaschia calocera*) is a beautiful rotter of wood that has started to turn up in many new places and in different habitats all over the world—most recently North America.

↙ The most infamous mushroom on the planet is the Death Cap (*Amanita phalloides*). This species is responsible for 90–95 percent of all mushroom fatalities globally. It is essential to familiarize yourself with this species if you collect wild mushrooms.

WILDFIRE THREATS

As the planet warms there are hotter and drier years, resulting in more frequent wildfires. Wildfires around the globe have caused tremendous loss to human life and property, and are inflicting lasting damage on species and ecosystems.

Of course, there are some ecosystems that have long been fire-prone and indeed various organisms require fire to thrive. For example, some conifers produce cones that remain on the tree and will not open until a hot fire passes through the area. Some forest fungi, too, do not seem to appear until after a fire.

FUNGI CAN HELP RESTORE POST-FIRE FORESTS

Just as fungi are a key component of healthy living forests, so they also play a key role in post-fire restoration. Of about 430 species of ascomycetes in the Pacific Northwest, more than 100 species require a forest fire to produce fruiting bodies. Many of these are quite small and easily go unnoticed.

"PYROPHILOUS" MUSHROOMS

Larger basidiomycete fungi also seem to fruit only after a fire. These include species of *Pholiota*, *Psathyrella*, *Inocybe*, *Tricholoma*, and *Clitocybe*, among other genera. Sometimes called phoenicoid fungi (for their ability to rise from the ashes like the phoenix of legend), pyrophilous (fire-loving) fungi are found all over the planet and on every continent except for Antarctica. Most are poorly known and only now coming under scrutiny, their crucial role in healthy forests being discovered.

BONFIRE CUPS AND BURN MORELS

One of the most-studied and important of the pyrophilous fungi is the little Stalked Bonfire Cup (*Geopyxis carbonaria*), a mycorrhizal symbiont of most forest conifers—although you're unlikely to see it except after a fire (see Chapter 3, page 44–45). This is usually the first mushroom to carpet a burned area in the spring following a wildfire. Bonfire Cups are seen as harbingers of the next fire mushrooms to emerge, the burn morels. Morels are big business, highly sought after, and thus much better known among pyrophilous fungi. *Pholiota highlandensis* is often the first gilled mushroom on the scene, right after the morels. Only recently was it discovered that, unlike most species of *Pholiota*, this fire-loving species lives as an endophyte partner within forest plants, but fruits only after devastating fire.

← *Pholiota highlandensis* is a post-fire pioneer species. This unassuming and drab-looking mushroom has a surprising lifestyle as an endophyte.

HABITAT LOSS

Our planet's ecosystems are becoming less diverse, less complex, and falling apart as, one by one, their constituent species are lost. A recent United Nations Summit on Biodiversity concluded that perhaps 1 million of the estimated 8.5 million species of plants, animals, and other organisms are in imminent danger of extinction. This loss of biodiversity seems to be accelerating and it's mostly due to habitat loss and overdevelopment.

ENDANGERED SPECIES

To date, very few fungi have been formally listed as threatened or endangered. Although many countries have a "Red List" for endangered fungi, most do not. Red List status signifies that things are very wrong in a habitat, calling for a need for study and preservation as immediate first steps. Possibly the most famous Red List fungus is *Pleurotus nebrodensis*, a critically endangered mushroom of northern Sicily.

REVERSING BIODIVERSITY LOSS

Loss of biodiversity is a serious stressor of the planet and action needs to be taken now. The path forward is clear: we must curtail overdevelopment and habitat loss, and we have to accelerate the ongoing survey of the planet's biodiversity. For species on the decline, we have to do our best to determine what's going on and turn that around. In many instances, the solutions may be complex, just as the reasons for an organism's decline may be complex.

RARE . . . OR MERELY RARELY SEEN?

It's tough to know when fungi are truly uncommon, or simply infrequently fruit and are uncommonly seen as a result. As an example, several closely related tooth fungi that live as tree saprobes are considered rare and threatened. The Tiered Tooth fungus, *Creolophus* (syn. *Hericium*) *cirrhatus*, Lion's Mane mushroom (*Hericium erinaceus*), and Coral Tooth fungus (*H. coralloides*) are all considered endangered in Europe (although the latter two are well known and quite common in North America). In the case of *Creolophus cirrhatus*, this mushroom is rarely seen and thought critically endangered. But recent studies of wood rot fungi, whereby samples of wood were taken from many sources and examined using molecular techniques, found that this mushroom was present pretty much everywhere but doesn't often create fruiting bodies. Thus, the mycelium is common throughout European forests, but rarely seen.

INVASIVE FUNGI AS PATHOGENS

Most of our commercially important plant pathogens are non-native fungi. All crop plants are at risk: cereals (Wheat Stem Rust), fruits (Peach Leaf Curl), tubers (Late Blight), and even lumber trees (White Pine Blister Rust).

Coffee Rust is not only destroying coffee plants wherever it goes, but also entire economies of nations that rely on this very economically important crop. Despite best efforts, Coffee Rust was bound to make its way to the New World. From Ceylon and India, the rust jumped to other countries in Asia and Africa and leaped the Atlantic to Brazil in the 1950s, then to Nicaragua in 1976. By 1981, *la roya*, as it's known in Spanish, had spread northward, arriving in Mexico.

Coffee Rust is caused by the rust fungus *Hemileia vastatrix*; most of this fungus's life cycle is still entirely unknown. Coffee Rust fungus is so widespread around the globe that there is no way to ever eradicate it. But through research using modern techniques—plus good old-fashioned smart cultivation practices—it is hoped that this economically important pest can be brought under control.

↓ White Pine Blister Rust, native to Asia but introduced to North America in the early 20th century, is among the most famous forest tree diseases.

↓ Sporulating orange pustules on the lower surface of a coffee leaf (a leaf stomate is visible top right). These are the sources of infectious spores.

→ Coffee (*Coffea arabica*) leaf showing symptoms of Coffee Rust (*Hemileia vastatrix*) infection on the leaf's upper surface. Coffee Rust pustules enlarge radially, producing new sori (clusters of spore-producing structures) near the leading edge of the chlorotic zone (where the leaves yellow).

MICROBES FOR A HEALTHY BODY—AND PLANET

You probably know that all organisms, including us, have a genome. This is the total genetic makeup of that organism. But did you know that we also have a microbiome? This is the community of microorganisms living on and inside eukaryotes like us. Mostly beneficial, these microbes are drivers of host health, enhancing immunity, nutrient acquisition, and tolerance to environmental stresses. Well, it follows that the microbiome of the entire planet is also a key driver of health—for the planet. And arguably, the most important microbes in any environment are the fungi.

CONSERVATION AND "REWILDING"

From this has emerged a "microbiome rewilding hypothesis" that suggests plant and animal health can be improved by reinstating key members of the diverse ancestral microbiota that were lost through domestication and industrialization processes, including changes

---

REWILDING OUR GUT MICROBIOME

Does rewilding the planet sound like science fiction? So did human fecal transplants just a few years ago—until they were tried. And they work! After a major surgical event or heavy use of antibiotics, our own microbiome—those beneficial microbes in our gut—can be depleted or even wiped out. Patients can become weak, lethargic, or even sickened by other bad microbes left unchecked by the missing good ones. To get things back on track, physicians sometimes prescribe a fecal transplant—literally a large capsule of feces (taken from a donor or more often from the patient prior to antibiotic treatment)—which is swallowed by the patient. Once inside the gut, the capsule dissolves and releases those healthy microbes to get things back in order.

---

in diet, plant and animal breeding, and the overuse of antibiotics, pesticides, and fertilizers. To date, however, it is unknown for most plant species whether (and which) microbial genera and functions were lost during plant domestication and to what extent rewilding can enhance the health and sustainability of modern crops.

~ Domestication of crop plants ~

Through stepwise processes, crop plants acquired all sorts of modern new traits, including larger seeds and fruits, determinate growth, photoperiod sensitivity, and reduced levels of bitter substances. Domestication was also accompanied by considerable habitat expansion and management practices, with increased reliance on external inputs like pesticides, fertilizers, and water to obtain higher yields. Naturally, the transition of plants from their native habitats to new ones led to changes in the microbiome composition of those plants. Most important to healthy plants are their mycorrhizal fungal partners and changes in plant microbiomes most impact these, with domesticated crops showing lower colonization and a decreased growth response to fungal symbionts, especially in fertilized soils.

REWILDING AND RESTORING OUR PLANET

Rewilding involves reinstating key ancestral microbes in agricultural soils or planting materials, and/or breeding modern crops with specific traits that support ancestral microbiota colonization. Once the ancestral forms of microbes are found (and "biobanked" for future use), it's time to do the rewilding—that is, to put them back in the environment. Rewilding approaches can offer a new avenue to harness the benefits of ancestral microbiota and do not preclude the use of domesticated crop cultivars or agricultural management practices, such as fertilizer. As rewilding research moves between the field and the lab, its value and integration in breeding programs for a new generation of "microbiome-assisted" crops await critical assessment in different agricultural settings.

FUNGI FASCINATE

Mushrooms have been a part of folklore, legends, and even celebrations around the world for a long time. This is especially so down through the ages in Europe. Mushrooms were considered strange—not quite animal, since they did not move, and not quite plant, because they did not photosynthesize. They often occurred in dark and damp, or otherwise putrid, places. Some were clearly poisonous. Where they came from and what they were doing in the environment was long a mystery.

MYSTERIOUS FAIRY RINGS

Fungi provide mystery, even today. You have likely seen mysterious green rings on grassy expanses of lawns and golf courses around your town. These "fairy rings" have been a source of fascination and myth for centuries to ancient Europeans and Native Americans of the Great Plains—and even to people today. Upon close inspection, you may see mushrooms emerging in the rings—some reaching full size overnight as if by some supernatural force. Fairies, elves, pixies, witches, dragons, and assorted amphibians have been blamed as well as celebrated for these manifestations.

Although fairy ring growth may still be mysterious to many people, mycologists have determined what is going on. A fungal spore arrives and germinates and, as fungal hyphal growth proceeds outward, the fungus digests the organic matter in the soil, including dead lawn thatch. This results in greener and taller grass where the ring of mycelium is growing. This is because the fungus excretes enzymes into the soil to digest the organic matter, but it doesn't reabsorb all the digested nutrition that results. Since fungi are messy eaters, the plants also get nitrogen and other nutrients released by fungal enzymatic action, and they, too, benefit. You won't always see the results of fairy ring fungi; in situations where the grass receives an abundance of water and fertilizer it will, of course, be unnaturally lush in growth, thus the physiological effects of the fungus will be mitigated.

THE MARCESCENT
FAIRY RING MUSHROOM

Many kinds of mushrooms will fruit in a fairy ring, probably the most ubiquitous worldwide being the Fairy Ring Mushroom (or Scotch Bonnet), *Marasmius oreades*. That *Marasmius* mushrooms can seemingly appear overnight is due to their marcescent habit—they can dry and wither, but then rehydrate when moisture returns. *Marasmius oreades* is not the simple saprobe it was once thought. Recent evidence suggests it is parasitic on the roots of grasses. In addition to cellulases and other enzymes, the fungus releases toxins, including hydrogen cyanide, which damage root tips, as well as impeding water percolation through the soil, all of which are detrimental to grasses.

THE MOST LEGENDARY MUSHROOM

Historians have deciphered and translated many nursery rhymes, fairy tales, and legends from all over the world that depict *Amanita muscaria*, the Fly Agaric mushroom, in a starring role. Scandinavian peoples have a number of legends about the Fly Agaric. Finns, Sami, and Laplanders collected this mushroom for its psychotropic and mind-altering properties.

The people of Siberia still use this mushroom for religious purposes, and this was documented firsthand in the 1980s by the late American mycologist Gary Lincoff and others who traveled there to investigate. Gordon Wasson and Valentina Pavlovna described the uses of this mushroom by Indigenous people of North America.

HOLIDAYS AND THE FLY AGARIC

Amanita muscaria is the very recognizable large mushroom with a red cap and white scales. Probably ever since Christmas trees were decorated during the holiday season, it has been a tradition in many countries to hang a Fly Agaric ornament on the tree; those made from blown glass in Germany and Czech Republic are works of art unto themselves. Probably centuries ago it was a real, dried mushroom that families would place on the tree.

SANTA . . . AND FLYING REINDEER?

Of the many myths with tie-ins to the Fly Agaric, in the West probably none is more famous (and repeated as if it were fact) than the notion that the modern-day Santa Claus (or Father Christmas, depending on your national origin) comes from Scandinavian or Siberian mushroom shamanism. It's plain as day that Santa's red and white costume represents the trippy mushroom with its red cap and white patches. Right? The flying reindeer, well . . . you hardly need to use your imagination for that one. It is well known that the reindeer, whitetail, and many other types of deer enjoy consuming this mushroom as well! Santa's sleigh flights represent one's spirit journey when under the psychotropic influence of the mushrooms. His large sack is full of gifts from the spirit world. That he shinnies down the chimney is easily connected with the way Siberian shamans would enter their yurts by way of the smoke hole at the top.

← It's long been a popular tradition in Europe to place a dried mushroom on Christmas trees and nowadays pretty blown-glass ornaments are even more popular.

MUSHROOM OF FABLES

A popular Swedish children's book called *Tomtebo Barnen* (Children of the Forest) depicts a little gnome family that lives in the woods. The children wear hats made from the red Fly Agaric (*Amanita muscaria*), which helps keep them safe in the forest while they are on their own. When danger approaches the children drop to the ground, revealing only their mushroom-cap heads, which blend into the forest floor—a sort of camouflage.

There seems to be a recurring theme of morbid curiosity with nursey rhymes and red mushrooms in Europe. Probably none is better known than the folktale of the Baba Yaga. She was part bogeyman, part witch, and depicted as a withered old crone. Many of the tales were meant to warn children from straying off into the woods, lest she grab them (or they fall victim to some other more realistic peril). Adults would point out that seeing a Fly Agaric in the woods was proof positive she was nearby or had just passed through.

↓ The Baba Yaga of European lore is usually depicted as an old crone, with or without a basket and broom, but usually with some red-capped Fly Agaric mushrooms nearby.

→ Mushrooms have been a part of folklore around the world for a long time. Mushrooms were strange and curious to people, and often depicted as all the more so in books and paintings. It's little wonder they have had a bad reputation as evil for most of history. Thankfully, that attitude is changing.

ALTERING THE COURSE
OF HISTORY

Fungi have changed the course of history many times. Plant pathogenic fungi have caused famines. Fungi have plagued humans and continue to extirpate some animals like amphibians and bats. Fungi have even been implicated in the deaths of some of the world's religious and political leaders down through the ages.

FUNGAL VICTIMS: EMPERORS,
POPES, AND KINGS

The rule of Pope Clement VII (1478–1534) is notable in the annals of history, not so much for its duration but for the world upheaval that happened during his reign, including The Reformation and the Sack of Rome. Clement VII was said to have been killed by eating Death Cap (*Amanita phalloides*) mushrooms, but some historians dismiss this theory. Austrian King Charles VI died in 1740 after eating a meal of Death Caps while on a hunting trip in the mountains. He led a lavish lifestyle and in death left the Hapsburg Empire in severe debt. Neither the royal family, nor his financial advisors, nor loyal subjects could stop him. But in the end, the mighty mushroom did.

~ Death to Caesar? ~

Possibly the most infamous of all murders attributed to poisonous mushrooms, and one that undoubtedly changed the course of world history, involved the death of the Roman ruler Claudius in CE 54. Presumably it was the fondness for *ovuli* that led to the highly prized edible *Amanita caesarea* being named Caesar's Mushroom. It also led to his death.

→ Caesar's Mushrooms are found throughout the Northern Hemisphere. The beautiful European species, *Amanita caesarea*, seen here, is a highly prized edible.

Claudius Caesar ascended the throne following the assassination of his nephew, Caligula. Following marriage to his fourth wife, Agrippina, Claudius adopted her son Nero. Most scholars believe the marriage was one of convenience and politically motivated. Even so, it lasted for many years and until the untimely death of Claudius. By all accounts Claudius was poisoned with his favorite dish of mushrooms—whether toxic amanita mushrooms were mixed with edible ones will never be known. What is clear is that Agrippina had repeatedly argued with Claudius to make her son, Nero, next in line to the throne. Claudius favored his son by blood, Britannicus. It's also clear that following the murder, Nero did become the Roman ruler . . . and we know how that turned out.

NECROTROPHS AND SARCOPHILES

Some fungi are "sarcophilous"—that is, keenly adapted to decompose animal carcasses as well as some nitrogenous animal wastes. Many of these fungi, including *Hebeloma*, are mycorrhizal. Sarcophilous fungi can often be stimulated to fruit by burying urea or other compounds that decompose into ammonia in the woods. Thus, in the absence of a fresh corpse, other sources of ammonia are a suitable habitat.

The corpse may hardly be cold when the Corpse Finder (*Hebeloma syrjense*) and Ghoul Fungus (*H. aminophilum*) set to work. It is entirely likely that spores of these poorly understood fungi are transported to corpses via flesh flies of the family Sarcophagidae, or other arthropods. The Corpse Finder is uncommon but occasionally found across the Northern Hemisphere, with the Ghoul Fungus known from Australia. All are understood to grow in association with decomposing animal remains and are sometimes mentioned in the literature for their utility in forensic science.

↙ Flies of the family Sarcophagidae (meaning "corpse-eating") specialize in finding carcasses. They unwittingly transmit sarcophilic fungi.

↙ The Ghoul Fungus is found only in Australia and in association with the decomposing remains of animals.

→ All life in the ecosystem is connected, and everything depends on everything else. Underpinning many of these connections are fungi, which are some of the key decomposers, pathogens, and symbionts of this world. Without death and decomposition, there could be no life.

WITCHERY

Most accounts have the hookah-smoking caterpillar in Lewis Carroll's *Alice's Adventures in Wonderland* sitting atop a big red mushroom with white patches. Indeed, *Alice in Wonderland* is one of the earliest depictions in literature of hallucinogenic mushrooms and their effects. Ideas about changes to perception of size—appearing larger (macroscopia) or smaller (microscopia) than in reality—is often the first thing that comes to mind when someone mentions this story. The idea for the story likely first came to Carroll by way of another mycologist: Mordecai Cooke. Lewis Carroll (his real name was Rev. Charles Dodgson) became enthralled with hallucinogenic fungi after reading an article by Cooke called "A Plain and Easy Account of the British Fungi" published in 1862 in the *Gardener's Chronicle*.

THE SORCERER'S APPRENTICE

Things got even more far-out for Mickey Mouse in the classic Disney animated movie *Fantasia*, originally a short film in 1940 based on Goethe's story *The Sorcerer's Apprentice* (set to the wonderful music of French composer Paul Dukas). While his master was away, the apprentice mouse did play. Poor Mickey tires of being mere apprentice to a sorcerer and tries his hand at magic, but soon finds himself unable to contain his sorcery. The film was later made into a full-length animated movie and employed many wonderful classical music components and gorgeous imagery, including dancing broomsticks, flowers, and Fly Agaric mushrooms doing the "Danse Chinois." Indeed, just about every animated movie (think Disney, Smurfs, Super Mario Bros., and so on) that depicts a woodland scene features stylized Fly Agarics.

ERGOT FUNGUS

There are more than 40 species of ergot, or *Claviceps,* fungi, all of which are parasites of grasses, rushes, and sedges. The best known is *C. purpurea*, found in all temperate regions and is a pathogen of cereal grains. The fungus produces a curved, purple-black sclerotium (ergot is the French word for "spur") on the host. Sclerotia, or ergots, are the overwintering stage of the fungus; during spring, spores are forcibly discharged into the air and infect the next cereal crop. Ergot fungi have a more sinister side: they produce toxic alkaloids. These mycotoxins are a health risk to humans and livestock alike. Modern cleaning methods remove ergots from grain before it is milled or used for animal feed, but the process is costly and may leave toxic residues.

~ Ergotism ~

In the Middle Ages, a frightening disease known as "St. Anthony's fire" would occur periodically. Symptoms included a tingling or burning sensation of the skin, paralysis, convulsions, tremors, and hallucinations. Ergotism, as it was called, can be lethal and thousands of people died during epidemics in the 1800s. Outbreaks occasionally happen to this day; the largest outbreak in modern times was in 1951 when an entire French village was afflicted. Long blamed on mass hysteria, the infamous Salem Witch trials of Massachusetts were likely a result of ergotism. Beginning in February 1692, some young girls in Salem fell into convulsive fits, screaming, and speaking in tongues. Hysteria swept through the village and townsfolk were accused of witchcraft. After trials lasting several weeks, 19 women were hanged and an elderly man crushed to death beneath heavy stones.

→ The cause of ergotism, *Claviceps purpurea*, growing on cereal grain. Arguably the most infamous of all ascomycete fungi on the planet, it causes much death and suffering.

PREDACIOUS FUNGI

Fungi have evolved all sorts of interesting lifestyles, ranging from mutualism to being pathogens of plants and decomposers of organic material. Likewise, the size, physiology, and chemistry of fungi can be quite disparate. While some fungi cause diseases of animals—including us—possibly the most interesting, and gruesome, are the fungi that are predators of animals. No, you need not look over your shoulder, we don't yet know of any fungi that hunt down vertebrates like humans. But invertebrates, be forewarned.

Insects, by far, represent the largest group of animals, so it's no surprise that pretty much all its members are commonly devoured by fungi: butterflies and moths, ants and bees, beetles, flies, and more. Other arthropods as well, including spiders, are frequently dined upon.

NEMATODES ON THE MENU

Nematodes comprise one of the largest groups of invertebrate animals. They are very small, round worms that can be found in just about any situation, although they are minute and mostly go unnoticed. There are many thousand named species and they range from saprobes to pathogens of terrestrial and aquatic plants and animals. Moreover, nematodes cause very serious economic losses in many of our agricultural crop plants. They also cause disease in livestock and some are gross pathogens of humans. Nematophagous fungi are found among the chytrids, zygomycetes, ascomycetes, and basidiomycetes. The latter include *Pleurotus* species, the Oyster mushrooms.

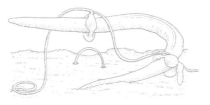

← A soil nematode moving through the soil is caught. Lasso Fungi are being studied and commercially deployed as an environmentally benign way to defeat a very serious pest.

HARNESSING FUNGAL POWER

There is an ever-growing demand for nematicides and worldwide sales exceed $1 billion; over the past decade, the market for natural biological nematicides has grown by 20 percent, which can be attributed to growing concerns about the environmental hazards posed by chemical controls and limited alternative crops for rotation. Native nematode-trapping soil fungi are obviously regarded as an ideal candidate for biological control and there are already two commercial preparations on the market.

~ Oyster mushrooms are saprobes—and predators ~

The specialized toxins and mechanisms to trap, kill, and ingest nematodes are just as diverse as the fungi themselves. Some, like *Pleurotus*, produce short, branching hyphae tipped with toxins that kill their prey on contact. Other fungi produce conidia that may be ingested or stick to nematodes as they swim past; upon germination the host is soon filled with fungal hyphae. Still other fungi produce swimming zoospores that are chemically attracted to nematodes, and hunt down and attach to their quarry, usually around an orifice.

LASSO FUNGI

Probably the most studied of nematophagous fungi are species of the genus *Arthrobotrys*, the Lasso Fungi. The hyphae of *Arthrobotrys* grow through the soil, like most other molds. But along the way, nematode traps are set. Some species of *Arthrobotrys* produce coils and loops of hyphae, coated with an adhesive that resembles a sticky net. The loops of some species act as snares—when a nematode attempts to swim through, the loop constricts quickly and tightly around the unsuspecting prey and it's caught fast. The constricting hyphal rings of *A. dactyloides* are formed by three cells; the sensation of a nematode passing through will trigger them. Once stimulated, the three cells rapidly inflate, severely constricting the nematode. Over about 24–36 hours, the interior of the nematode is completely filled with hyphae and then fully digested from the inside out.

THE LARGEST FUNGI

Everyone always wants to know what the largest mushroom is and where it may be found. The largest known gilled mushroom, aptly named the Titan Mushroom, is *Termitomyces titanicus*. The stalk of this behemoth may be several feet long and sports a massive cap more than 3 ft (1 m) wide. A true titan! And throughout their range, *Termitomyces* species are prized edibles and widely sold in markets.

Termitomyces species are obligate biotrophs of termites, farming them within their subterranean nests. Although termites are found throughout the world, including Australia, where they build massive nests, *Termitomyces* are found only in Africa and Southeast Asia. All termites eat plant matter but rely on microbes living within their gut to digest the cellulose for them. The termites eat fresh plant material, which passes through their intestine and is molded together to form a substrate for the fungi cultivated deep in the insects' labyrinthine nest. Some termites rely exclusively on mycelia as food, while other species ingest this material, which affords them the enzymes to digest the cellulosic matter they forage.

↓ A welcome find, *Termitomyces titanicus* is a popular edible mushroom in Africa and Southeast Asia. The fruiting body can grow to an astounding size, with a stalk several feet long.

→ How do *Termitomyces* species "escape" their host's gardens deep underground? In some cases, fungal mycelium is carried from the nest by termites leaving the nest to start their own colonies. At other times, *Termitomyces* produce mushrooms with very long, "root-like" stems that break through the soil surface. The fruiting body is able to penetrate the wall of the nest and compacted soil, then emerge above ground.

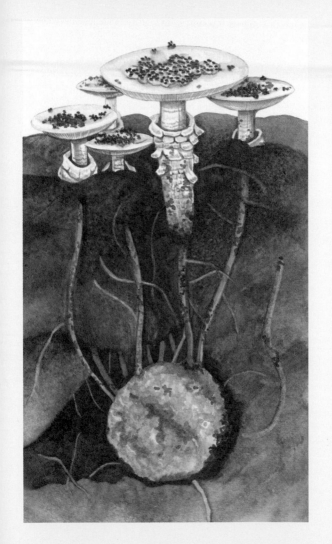

THE TINIEST FUNGI

Fungi are considered microbes, thus much of the time these organisms are so small that a microscope is required to see them. Yeasts are very small, single-celled, free-living fungi. But the smallest fungi are even smaller.

Microsporidia are the smallest known fungi. In fact, they are so small that they are unable to live on their own. Instead, they are obligate intracellular parasites of other organisms. That's right, they must live within the cells of animals, mostly insects like honeybees, but a few are known parasites of humans.

It's unlikely you will spy a microsporidian on your next foray into the woods. The entire life of a microsporidian, including replication, takes place within the cell of its host. If they came from a true fungal ancestor, they long ago gave up hyphal growth to live as endosymbionts. Microsporidians are some of the smallest known eukaryotes and have the smallest eukaryotic genomes.

↓ Microsporidia (*Nosema* sp.) are tiny, unicellular parasites and cause a commercially important disease of honey bees.

↓ How tiny? The spores of the fungus *Nosema apis*, a bee pathogen, are shown here among plant pollen.

→ *Sordaria macrospora*, an ascomycete decomposer, creates very tiny fruiting bodies that look somewhat like puffballs. However, their spores are produced in tube-like asci and are released by a squirt gun-like mechanism.

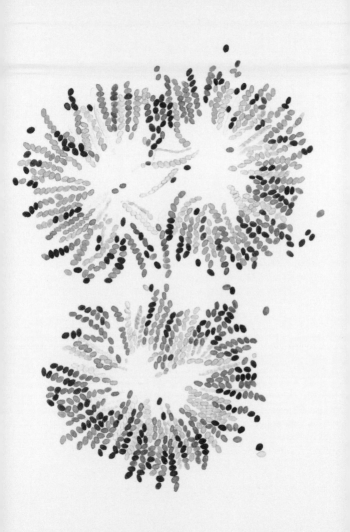

FUNGI IN UNUSUAL ENVIRONMENTS

I t would likely come as a surprise to find fungi in extremely hot and dry deserts. Likewise, you might not expect to find fungi above the tree line on mountain peaks, or in the frozen wastes of Antarctica. And yet many fungi are keenly adapted to these harsh habitats.

AQUATIC MUSHROOMS

While we know of marine and freshwater fungi, mushroom-producing fungi are a real anomaly. Indeed, mycologists still don't know much about their lifestyle. While several ascomycetes are known to grow and fruit under water, no aquatic gilled mushrooms had previously been known until *Psathyrella aquatica* was discovered in Oregon's Rogue River in 2005. This mushroom looks similar to most other species of the genus that you would expect to find in the woods or compost piles near your home.

~ Underwater sporulation ~

How this mushroom produces spores is still not known. Ballistospory is not supposed to work under water. This mushroom must somehow create air bubbles on the gill surfaces and fire spores into them; rafts of spores have been documented floating in the vicinity of fruiting bodies. Alternatively, spores might also be spread as the mushrooms wither and float away.

↗ Ballistospory is not supposed to work under water but somehow the bizarre *Psathyrella aquatica* has figured out a way. Scientists are still unsure how it achieves this.

LIFE UNDERWATER POSES CHALLENGES

Everything about the lifestyle of the *Psathyrella aquatica* mushroom raises questions. Probably the most obvious is: how are its spores transmitted upstream? Some dispersal vector must be required to counter the constant flow of water downstream; thus, it is suspected that aquatic insects are involved. Oregon-based mycologist Jonathan Frank braved the icy waters to film invertebrates associated with the underwater mushrooms. Upon capturing and dissecting these insects, spores of *Psathyrella aquatica* were observed in the guts of caddisfly, mayfly, and black flies. These observations suggest that aquatic insects are involved in spore dispersal, either as mycophagists, grazers, or filter-feeders, collecting spores as they move under water along the mushrooms. While more data will be needed to confirm the roles of these invertebrates, aquatic insects have the ability to counter the flow of water downstream and potentially move spores upstream to suitable new habitats. Additionally, the insects may be consumed by fish or birds and move spores even farther.

TOXIC FUNGI

While most fungi are harmless to us and other animals, some are certainly poisonous enough to make us sick—in fact, some can be lethal. Some molds that contaminate our agriculture products and foods (there could be some lurking in your refrigerator!) can produce toxins. Most notorious, however, are some of the world's wild mushrooms.

WITHOUT WARNING

If you are at all interested in collecting wild mushrooms for food, you absolutely must be familiar with the potential dangers of consuming poisonous species. Dangerous mushrooms oftentimes resemble other familiar edible mushrooms, even cultivated species. Unfamiliar pickers erroneously assume that poisonous mushrooms will warn of impending danger with garish colors, foul odors, or bitter or off-putting tastes. In nature, most toxic or venomous organisms do display aposematic or "warning colorations"—the bright red or yellow colors, for example, of many animals and plants. Fungi do not follow these rules of nature. The most commonly encountered poisonous mushrooms are drably colored browns or grays, and many are pure white. Furthermore, most taste quite pleasant—there's nothing to warn you that what you're currently savoring in a prepared dish is about to kill you.

→ Are you willing to risk your life? The deadly Death Cap mushroom is very similar in appearance to popular edibles like the western North America Caesar Amanita (*Amanita vernicoccora*). Both species are shown here.

MOST TOXIC MUSHROOMS

Amanita phalloides is one of the most widespread mushroom species worldwide. Known as the Death Cap, it was first described in Europe and is now known in all continents except Antarctica. This mushroom is deadly poisonous and responsible for the vast majority of mushroom poisoning deaths (90–95 percent) worldwide. It has made the headlines in Australia following a number of deaths, including patrons who'd consumed wild harvested mushrooms at a restaurant a few years ago. Europeans are more wary of this species because it's always been there.

~ Becoming more widespread ~

That *Amanita phalloides* is now so widespread is attributed to its ability to pair up with a wide assortment of host trees, including horticultural and economically important species. Thus, it is frequently transported and transplanted globally. In North America, the Death Cap is mostly associated with oaks on the West and East Coasts. In just a few decades' time, the range of this mushroom has expanded dramatically and there is no reason to think that it won't continue to grow.

FUNGAL MIMICRY

W hen it is time to reproduce, fungi produce spores and most launch them into the air. Once they alight and germinate, a new generation of that species is underway. Some fungi do sporulation differently. Fungi may go to great lengths and have evolved amazing tricks to coerce animals into transmitting their spores for them. Some fungi produce foul odors to entice insects that feed on detritus or carrion.

Mimicry is the adaptive resemblance of one organism to another. Of course, our best known examples of mimicry are from animals that exploit one another to gain protection from predators—as seen in Viceroy butterflies mimicking the toxic Monarch butterfly. But there are equally fascinating examples of mimicry among plants and fungi, and no doubt many still awaiting discovery. Some fungi mimic flowering plants—even producing "pseudoflowers" that dupe pollinators into disseminating their spores. Through natural selection, these fungal tricksters beat the plants at their own game.

PSEUDOFLOWERS AND FALSE FRUITS

That flowering plants compensate animals (like insects, birds, and mammals) to vector pollen and seeds is, of course, well known. Flowering plants likely offer a reward to the pollinator to prevent damage to the ovary or other floral tissues; likewise, fleshy fruits offer a nutritious reward to an animal vector that transmits seeds. Amazingly, fungi may mimic flowers and fruits to the same end. Some fungi that infect plants induce them to produce fruit-like ("false fruits") or flower-like ("pseudoflowers") structures to entice animals to move their spores around.

The leaves and floral shoots of blueberry and huckleberry plants frequently become parasitized by the discomycete fungus *Monilinia vaccinii-corymbosi*. Once inside, the fungus produces hormones that induce infected leaves to change in appearance, resembling the plant's own flowers and enticing pollinators to come and visit. On the surface of these pseudoflowers, the fungal hyphae seem to produce sweet exudates, along with infectious conidia; while the pollinators are sipping the fungal "nectar," they are also picking up fungal spores that they vector to healthy flowers, spreading this economically important plant pathogen. The fungus overwinters as sclerotia in "mummified" fruits (hence the disease name: mummyberry) on the soil. The disease begins in the spring when the sclerotia produce apothecia that release ascospores to newly emerging leaves—the leaves that will become the pseudoflowers for this floral mimic.

TURNING THE TABLES ON FUNGI

There are very few examples of "pseudopollination," where plants turn the tables and mimic fungi. Though one plant is so good at mimicking a fungus that it's unlikely anyone (even an expert botanist) would notice! Within the cloud forests of the Central and South American tropic region is an orchid known as *Dracula*. In fact, there are more than 100 species of *Dracula* orchids whose showy flowers look—and smell—like mushrooms. *Dracula* species inhabit sodden, drippy ledges where few other flowering plants dare tread. As a result, there are few pollinators to be found. But there are plenty of mushrooms fruiting year-round from the moist humus. So, *Dracula* species entice mushroom-feeding flies to satisfy the need for pollination. Floral parts closely resemble gilled mushrooms, complete with mushroom smell. That's right, the same odor of mushrooms, a chemical called 1-octen-3-ol, is produced by the orchids, thus completing the charade.

← The name *Dracula* literally means "little dragon" and was applied to the genus because of the blood-red-colored flowers and long, sinister-looking sepal spurs.

GLOSSARY

anamorph
The asexual state or form of a fungus. Compare with teleomorph.

ascocarp
A fruiting body containing asci and ascospores.

ascomycetes
A group of fungi that reproduce sexually by the endogenous formation of ascospores in an ascus.

ascospore
A haploid spore produced within an ascus following karyogamy and meiosis.

ascus (pl. asci)
A sac-like chamber that produces ascospores; asci are characteristic of the Ascomycota.

aseptate
Lacking septa, often pertaining to the hyphae seen in zygomycetes (see also coenocytic).

basidiocarp
A fruiting body bearing basidia and basidiospores.

basidiomycetes
A group of fungi that reproduce sexually by producing basidiospores from a basidium.

basidiospore
A haploid spore produced on a basidium following karyogamy and meiosis.

basidium (pl. basidia)
A club-shaped chamber that produces basidiospores; basidia are characteristic of the Basidiomycota.

biotrophic
Adjective describing an organism that lives and multiplies on another.

coprophilous
Growing in or on dung.

fruiting body
Also termed "mushroom," this is the sexual spore-producing structure of ascomycete or basidiomycete fungi. Authors may also say "fruitbody" or "fruit body."

fungi imperfecti
An informal and polyphyletic grouping of unrelated fungi that are known only by their anamorphic (asexually reproducing) forms. Many of these are the anamorphs of ascomycetes and basidiomycetes, but without sexual fruiting bodies their affinities remain obscure.

gills
The lamellae, or gill-like hymenial structures of agaric mushrooms.

haustorium (pl. haustoria)
A specialized appendage of a parasitic fungus that penetrates the host's tissues, but does not penetrate the host's cell membranes; haustoria of arbuscular fungi are called "arbuscules."

hymenium
The fertile tissue giving rise to and bearing the sexual spores (e.g., the gills of agarics and the pores of boletes and polypores).

hymenophore
Structure bearing the hymenium, the mushroom.

hypha (pl. hyphae)
A single filament of a fungus.

mycelium (pl. mycelia)
The mass of hyphae making up the thallus of a fungus.

mycobiont
The thallus-producing fungal partner in the symbiotic associations known as lichens.

mycosis
Fungal disease of humans.

nonseptate
Lacking septa; also termed "aseptate."

photobiont
The photosynthesizing algal or cyanobacterial partner in the symbiotic associations known as lichens.

rhizomorph
A mycelial strand of aggregated parallel hyphae attached to the basal portion of some mushrooms.

saprobic
Obtaining nourishment from dead or decaying organisms.

saprotrophic
Adjective describing an organism that feeds on dead organic matter.

sclerotium (pl. sclerotia)
A highly condensed mass of undifferentiated sterile (asexual) hyphae typically encased in a hard, woody, thick, dark rind. Such structures enable those fungi that produce them to survive under adverse environmental conditions.

septum (pl. septa)
A "partition," or cross-wall, in a hypha, cell, or spore.

sterigma (pl. sterigmata)
A small, narrow, stalk-like structure at the apex of a basidium upon which a basidiospore forms.

stroma (pl. stromata)
A compact mass of fungal tissue on or within which fruiting bodies develop.

taxonomic
Adjective referring to the classification and/or nomenclature of an organism or group of organisms.

taxonomy
The discipline devoted to the collection, cataloguing, classification, and naming of organisms.

teleomorph
The sexual stage of a fungus. Compare with anamorph.

zygospore
A thick-walled sexual spore formed by the fusion of two similar gametangia; characteristic of the zygomycetes.

FURTHER READING

Ainsworth, G. C. 1976. *Introduction to the History of Mycology*. Cambridge University Press, Cambridge; 359 pp.

Alexopoulos, C. J., C. W. Mims, and M. M. Blackwell. 1996. *Introductory Mycology*, 4th edition. Wiley, New York; 869 pp.

Arora, D. 1986. *Mushrooms Demystified: A Comprehensive Guide to the Fleshy Fungi*, 2nd edition. Ten Speed Press, Berkeley; 959 pp.

Benjamin, D. R. 1995. *Mushrooms: Poisons and Panaceas*. W. H. Freeman and Company, New York; 422 pp.

Boughler, N. L., and K. Syme. 1998. *Fungi of Southern Australia*. University of Western Australia Press, Nedlands, WA, Australia; 391 pp.

Bunyard, B. A. 2022. *The Lives of Fungi: A Natural History of Our Planet's Decomposers*. Princeton University Press, NY and London, 288 pp.

Bunyard, B. A., and T. Lynch. 2020. *The Beginner's Guide to Mushrooms: Everything You Need to Know, from Foraging to Cultivation*. Quarry Books, Beverly, MA; 160 pp.

Dugan, F. M. 2008. *Fungi in the Ancient World: How Mushrooms, Mildews, Molds, and Yeast Shaped the Early Civilizations of Europe, the Mediterranean, and the Near East*. APS Press, St. Paul; 140 pp.

Harding, P. 2008. *Mushroom Miscellany*. Collins, London; 208 pp.

Hudler, G. W. 1998. *Magical Mushrooms, Mischievous Molds*. Princeton University Press, New Jersey; 248 pp.

Kendrick, B. 1992. *The Fifth Kingdom*. Focus Publishing, Newburyport, MA; 386 pp.

Laessøe, T., and J. H. Petersen. 2019. *Fungi of Temperate Europe*. Princeton University Press, New Jersey; 1708 pp.

Letcher, A. 2007. *Shroom: A Cultural History of the Magic Mushroom*. Harper Collins, New York; 360 pp.

Lincoff, G. 1981. *National Audubon Society Field Guide to Mushrooms*. Knopf, New York; 926 pp.

Marley, G. A. 2010. *Chanterelle Dreams, Amanita Nightmares*. Chelsea Green Publishing, Vermont; 255 pp.

McIlvaine, C. 1900. *One Thousand American Fungi*. Bobbs-Merrill Company, Indianapolis; 749 pp.

Millman, L. 2019. *Fungipedia: A Brief Compendium of Mushroom Lore*. Princeton University Press, New Jersey; 200 pp.

Petersen, J. H. 2012. *The Kingdom of Fungi.* Princeton University Press, New Jersey; 265 pp.

Phillips, R. 2010. *Mushrooms and Other Fungi of North America.* Firefly Books, New York; 319 pp.

Ramsbottom, J. 1953. *Mushrooms & Toadstools: A Study of the Activities of Fungi.* Collins, London; 306 pp.

Rolfe, R. T., and F. W. Rolfe. 1925. *The Romance of the Fungus World: An Account of Fungus Life in Its Numerous Guises, Both Real and Imaginary.* Lippincott Co., Philadelphia; 308 pp.

Schaechter, E. 1997. *In the Company of Mushrooms.* Harvard University Press; 296 pp.

Taylor, T. N., M. Krings, and E. L. Taylor. 2015. *Fossil Fungi.* Academic Press, London; 382 pp.

Webster, J., and R. Weber. 2007. *Introduction to Fungi,* 3rd edition. Cambridge University Press, Cambridge; 841 pp.

Other Resources

Associazione Micologica Bresadola: ambbresadola.it

Australasian Mycological Society: australasianmycologicalsociety.com

European mushroom information: fungus.org.uk

European Mycological Association: euromould.org

Fungal Network of New Zealand and New Zealand Mycological Society: funnz.org.nz

Fungi Magazine: fungimag.com

Fungi of California: mykoweb.com

Index Fungorum: indexfungorum.org

Mushroom Expert: mushroomexpert.com

Mushroom Observer: mushroomobserver.org

North American Mycological Association: namyco.org

INDEX

aflatoxin 89
Amanita caesarea 136
    muscaria
        104, 115, 120, 132–4
    phalloides 120, 121, 151
    vernicoccora 150
amber 46, 47
ambrosia beetles 78
amphibians 64–5
animals 10
    fungal diseases
        64–5, 111
    rewilding 128–9
Antarctic fungi 38–9
antibiotics 110
aquatic mushrooms 148–9
arbuscular mycorrhizal fungi
    (AMF) 84–5, 118–19
Armillaria 17, 35, 60, 68, 69
Arthrobotrys 143
Artillery Fungi 40
ascomycetes 18, 48, 70
asexual fungi 20, 76–7
Ash Bolete 102–3
Aspergillus
    21, 88, 89, 106–7, 111
assassinations 136–7

bark beetles 79
basidiomycetes
    16–17, 48, 70
bats 64, 65
Battarrea 36
biodiversity loss 124
bioluminescence 68–9
Birch Polypore 25
bird's nest fungi 74
blights 32–3, 76–7, 126
Bonfire Cups 45, 123
Botrytis cinerea 107
bracket fungi 24–5
brewing 106
brown rot fungi 60
butt rots 40

Caesar's Mushroom 136
Cedar Apple Rust 32
Ceratocystis 95

cheese production 107
chytridiomycosis 64, 65
chytrids 12, 13
citizen science 98–9
Cladosporium 39
classification
    13, 16, 20–1, 48, 96
Claviceps purpurea
    33, 141
climate change 118–119
coenocytic fungi 12
Coffee Rust 126
collections 97
coprophilous fungi 38, 66
Cordyceps 115
Corn Smut 33
Cordyceps 115
Corpse Finder 138
Crown Rust 57
cryptic fungi 96–7
cultivation 100
cup fungi 18
Cyttaria 54

Death Cap 120, 121, 151
decomposition 58–9
    necrotrophs and
        sarcophiles 138
Deconica coprophila 66
desert fungi 36
    Desert Shaggy
        Mane 36
Dictyophora 28
discovering new species
    34–5, 96–7
DNA analysis
    20–1, 96, 98–9
DNA barcoding 99
Downy Mildew 94
drugs 90, 110–12
Dry Rot 43
Dutch Elm Disease 95

ectomycorrhizal fungi
    (EcM) 83, 84
edible fungi 100, 116, 136
Elaphocordyceps 31
endangered species
    103, 124–5

endomycorrhizal fungi 84–5
endophytic fungi 35, 89
epiphytic fungi 35
ergot 33, 141
    ergotism 141
eukaryotes 10–11, 146
evolution 46–8
    co-evolution 54
    convergent evolution
        50–1
extinction 102–3

fairy rings 130
    Fairy Ring Mushroom
        131
fermentation 106
fire fungi 44–5, 123
Fly Agaric
    104, 115, 120, 140
    mythology 132–4
food production 106–7
fossil fuels 59
fossil record 46–7, 82
fungal damage 42–3
fungi 8–9, 10–11
    classification
        13, 16, 20–1, 48, 96
    fungal groups 48
    individual fungus 11
    named species 12, 97
    specialists 14
    tiniest 15
fungus-like organisms 11
Fusarium 88–9

Ganoderma 39, 40, 60
Ghost Pipes 92–3
Ghoul Fungus 138
glomalin 119
Golden Oyster mushroom
    120, 121

habitats 34–9
    fire fungi 44–5
    habitat loss 124
    our homes 42–3
    underwater 148–9
    urban areas 40

hallucinogenic
  mushrooms 112, 140
Hemileia vastatrix 126
honey bees 146

Impudent Stinkhorn 28
Inonotus 40, 60
insect associates 35
invasive species
  104, 120–1
  plant pathogens 126
IUCN (International Union
  for Conservation of
  Nature) 103
  Red List 124

Laetiporus 22, 40, 60
Lasso Fungi 143
Late Blight 76–7, 126
lichens 38–9, 86

medical uses 110–12
melanins 38, 39
microbiomes 128–9
microhabitats 34–5
microsporidia 146
migration 54
mimicry 80, 152–3
morels 18, 123
mushrooms 16
  caps 16
  form and function 22
  mycelium 17
  spores 17
  stalks 16
  world's largest 144
mycelium 17, 83
Mycena 68
mycology 94–5
mycorrhizal fungi
  48, 82–3, 92–3
mycoses 111
mycotourism 116
mycotoxins 88–9

Nameko 100
necrotrophs 138
nematicides 143
nematode trappers 142–3
Noble Rot 107
Nosema apis 146
obligate associations
  82, 144

Omphalotus 68
Onygena 62, 63
Orange Pore fungus 121
orchids 93, 153
Ötzi 25

Panellus 68, 69
Peach Leaf Curl 126
Penicillium
  39, 88, 107, 111
  penicillin 110
Phaeolus schweinitzii
  40, 58, 60
*Phanerochaete
  chrysosporium* 60
Pholiota 44, 123
Phytophthora infestans
  76–7
Pilobolus 14, 66
plants 10, 46, 92–3
  fungal mimicry 152–3
  plant pathogens 126
  rewilding 128–9
Pleurotus
  60, 100, 124, 142–3
  caryopilites 120, 121
Podaxis 36
polypores 24–5, 40
popular culture
  114–15, 140
protists 10, 11
Psathyrella 123
  aquatica 148–9
Psilocybe 112
psychedelics 112
puffballs 26–7
Purple Fairy Club 51
pyrophilous fungi
  44–5, 123

reproduction 70–1
  asexual oomycetes 76–7
  see spores
rewilding 128–9
ringworm 33
Robigalias 57
rusts 32–3, 56–7, 126

sarcophiles 138
Scarlet Berry Truffle 80
*Schizophyllum
  commune* 71
Shiitake 60, 100

smuts 32–3
Sordaria macrospora
  146
spores 17
  active spore dispersal
    78, 80
  ascospores 18
  Buller's drop 72
  passive spore dispersal 74
  spore production 72–3
*Sporophagomyces
  chrysostomus* 25
statins 111
stinkhorns 28
Stonemaker Fungus 45
Sudden Oak Death
  (SOD) 77
sustainable myco
  materials 108–9

Telluride Mushroom
  Festival, USA 116
Terfezia 36
Termania 36
Termitomyces 144
Tinder Polypore 24, 25
Titan Mushroom 144
toxic fungi 150–1
  mycotoxins 88–9
Trichoderma reesei 43
truffles 30, 36, 80
  false truffles 52–3
  truffle parasites 31
Tulastoma 36

urban areas 40

Wheat Stem Rust
  56, 57, 126
White Pine Blister Rust
  56, 126
white rot fungi 60, 79
white-nose syndrome
  (WNS) 65
wildfires 122–3
wine production 107
wood decomposition 59
wood rot 60
wood wasps 79
wood wide web 83

zombie fungi 115
zoochory 78–9, 80

159

ACKNOWLEDGMENTS

My tremendous thanks to the photographers and illustrators (see image credits, page 4) and to the entire team who made this book beautiful, and who were fun to work with: Ruth Patrick, Lindsey Johns, Caroline West, Slav Todorov, Tugce Okay, and Ian Durneen. Thanks also to the many authors of articles published in *Fungi Magazine* over the years, some of which were catalysts for features in this book. Many thanks to the publisher Nigel Browning, UniPress, and the Princeton University Press for the opportunity to write more stories about those weird and wonderful fungi.

ABOUT THE AUTHOR

Britt Bunyard, PhD, is the founder, Publisher, and Editor-in-Chief of the mycology journal *Fungi*, in print since 2008. Bunyard is a former university professor and has published over 100 academic and popular science papers. He has collected fungi and lectured throughout North and South America, Europe, and Asia; annually, he leads mycological expeditions throughout the world. One such expedition was the subject of a documentary film "Look Down Not Up" (2022), produced by documentary filmmakers Alok Siddhi Tuladhar and Dusty Shiva Panthi of Kathmandu, Nepal. Britt has authored several books, including *The Lives of Fungi* (Princeton University Press, 2022), *The Beginner's Guide to Mushrooms* (Quarry Books, 2021), *Amanitas of North America* (The Fungi Press, 2020), and *Mushrooms and Macrofungi of Ohio and Midwestern States* (The Ohio State University Press, 2012). Britt has served as Executive Director of the Telluride Mushroom Festival since 2014. In 2021 he was awarded the Gary Lincoff Award "For Contributions to Amateur Mycology," by the North American Mycological Association—NAMA's most prestigious honor for American mycologists.